CAMBRIDGE MONOGRAPHS ON MECHANICS AND APPLIED MATHEMATICS

FREE OSCILLATIONS OF THE EARTH

FREE OSCILLATIONS OF THE EARTH

E. R. LAPWOOD

Emeritus Reader in Theoretical Seismology, University of Cambridge

T. USAMI

Professor, Earthquake Research Institute, University of Tokyo

CAMBRIDGE UNIVERSITY PRESS

CAMBRIDGE

LONDON · NEW YORK · NEW ROCHELLE

MELBOURNE · SYDNEY

Published by the Press Syndicate of the University of Cambridge
The Pitt Building, Trumpington Street, Cambridge CB2 1RP
32 East 57th Street, New York, NY 10022, USA
296 Beaconsfield Parade, Middle Park, Melbourne 3206, Australia

First published 1981

Printed in Great Britain at the University Press, Cambridge

Photosetting by Thomson Press (I) Ltd., New Delhi

British Library Cataloguing in Publication Data

Lapwood, Ernest Ralph
Free oscillations of the earth. –
(Cambridge monographs on mechanics and applied mathematics).
1. Free earth oscillations
I. Title II. Usami, T III. Series
551.2′2 QE539 80–40705

ISBN 0 521 23536 7

TO
SIR HAROLD JEFFREYS
AND
DOCTOR KIYOO WADATI

The four terrible things: earthquake, thunder, fire and father.

Japanese proverb

Underneath the house
Of the millionaire also
Lies the earthquake belt.

Haiku by Ichigen

When an earthquake strikes, flee to the bamboo forest.

Japanese proverb

The giant crab deep under the ground swings its huge claws and grips the tail of the great eel in a surprise attack; the eel, writhing in unbearable pain, causes the earthquake.

Legend in Okinawa

CONTENTS

PREFACE

During the past seventy years the interplay of scientific theory and observation has been most beautifully exemplified in the investigation of the normal modes of vibration of the Earth. When a great earthquake strikes, the Earth's deepest modes are set in motion, and because of their long periods they lose energy only slowly. The Earth rings like a great bell for many days in these grave modes. At the same time a long sequence of overtones is energised; these involve strains of only about one part in a thousand million, but they can be detected and identified by modern seismographs and gravimeters.

The mathematical theory of such vibrations already existed seventy years ago, but its application to the determination of the characteristic frequencies of a realistic Earth-model could not be completed until high-speed electronic computers arrived to deal with the extensive calculations that were needed.

The theory and observations that have accumulated during these twenty years are well known to seismologists, but are to be found only in original research publications and review articles, or in brief summaries as small parts of books on seismology or geophysics. We think that there is a need for a systematic account of the subject, which can form an introduction for graduate students and a convenient reference for others. This book aims to meet that need.

Our general method of exposition in the following chapters is to introduce simpler models first, and to add complicating factors step by step. We choose this method, which follows the historical development, rather than that of starting from a most general system and treating simpler models as particular cases, because most of the ideas and methods which we introduce can be better grasped and explored when freed from more advanced or complex developments. Thus we begin in Chapter 2 with a uniform sphere, perfectly elastic and isotropic, with neither gravity nor rotation. We do not introduce the effect of gravity until Chapter 5, of rotation until Chapter 7. The first eight chapters omit consideration of the excitation problem, that is, of the manner in which vibrations are set up. That problem comes in Chapter 9, after all the relevant theory of free oscillations has been displayed.

Our primary aim has been to describe basic theory and results. However, in Chapter 10 we survey current work briefly, and give references to the intricate studies of the effects of lateral heterogeneity, anisotropy and anelasticity for which a book of this length can find no place.

We have made it our aim to present, by means of simple examples and analogies, the physical principles and interpretations involved at each step. Where we have been forced to omit mathematical detail we have tried to make the conclusions plausible by heuristic arguments, while supplying references where the rigorous treatment can be studied. We have included many diagrams that embody the results of computations for simple Earth-models, since these show clearly the patterns of eigenfrequencies, the forms of eigenfunctions, and intriguing problems still to be solved.

We assume that the reader has such knowledge of the following subjects as should have been acquired in an undergraduate course in Mathematics or Theoretical Physics: vector analysis, Cartesian tensors, differential equations, functions of a complex variable, special functions of Mathematical Physics, theory of vibrations and of wave propagation (including group-velocity and dispersion). We also assume a knowledge of the elements of seismology, such as may be found in Bullen's *An Introduction to the Theory of Seismology* or Jeffreys's *The Earth* (see references). This should include elementary classical elasticity, P and S waves, Rayleigh and Love waves, refraction at a plane interface, internal structure of the Earth.

We have included a fairly extensive list of references, which should enable the reader to find detailed discussion of all topics we have treated, and to proceed to recent papers that lie at the boundary where current research is active.

We are very grateful to colleagues for help by criticism, comment and suggestion; in particular to Dr T. Odaka, Miss Beverley Moore, Professor Alan Cook, Dr John Hudson and Dr Brian Kennett.

1

INTRODUCTION

1.1 Kelvin's estimate of frequency of a vibration of the whole Earth

In the middle of the nineteenth century many scientists believed that the Earth was a molten ball with a thin crust of solid rock. As evidence they could point to the rapid increase of temperature with depth from the surface (10–50 °C per km) and the flow of molten lava in volcanic eruptions. Against this assumption Lord Kelvin quoted what he described as 'the grand idea of learning the physical condition of the interior from phenomena of rotary motion presented by the surface' (Thomson, 1863a). He claimed that the observed precession and nutation of the Earth's axis implied that the spinning Earth behaved dynamically more like a rigid body than a fluid ball. It could not, of course, be perfectly rigid; Kelvin therefore attacked the problem of obtaining an estimate of its overall rigidity (*rigidity* being defined as the ratio of shearing stress to shear strain in an elastic body). He argued as follows: the gravitational attraction of the Moon and Sun raises ocean tides, and must also raise corresponding bodily tides in the solid Earth. The *observed* ocean tide is the difference between these. The semi-diurnal tide is the most prominent, but the fortnightly tide is the most appropriate for such a study, since its long period makes possible the use of statical rather than dynamical equations. Kelvin in 1863 calculated on the one hand the response of a sphere of perfect liquid to a tide-raising potential, and on the other hand the response of a perfectly elastic sphere of given rigidity. He was able to conclude that if the elastic sphere had the rigidity of glass the height of the bodily tide would be about $\frac{3}{5}$ of the theoretical liquid tide, while if it had the rigidity of steel the corresponding ratio would be $\frac{1}{3}$. Thus for a glass sphere the *observed* liquid tide would be $\frac{2}{5}$ $(= 1 - \frac{3}{5})$ of the theoretical tide, and for steel $\frac{2}{3}$ (Thomson, 1863b).

Tidal measurements showed that the observed fortnightly tide was in fact about $\frac{2}{3}$ of the theoretical tide, and Kelvin concluded that the Earth behaved, in responding to long-period tidal forces, as an elastic sphere as rigid as steel. He requested and obtained from his brother, a fellow physicist, a measurement of the speed of shear waves in steel. Using this he found a time of 68 min for a shear wave to pass along a diameter of uniform elastic sphere of the size of the Earth and the rigidity of steel. Arguing that this gave the right order of magnitude for an elastic vibration of the whole Earth, he had thus obtained an estimated eigenfrequency for the first time. This was the dawn of the study of the free oscillations of the Earth, which is the subject of this book.

1.2 General properties of small oscillations of a finite body

Before we advance further into the history of that study, it may be instructive to describe the general characteristics of the free oscillations of a finite medium. Any bounded flexible or elastic body can vibrate freely in one of an infinite set of characteristic modes. For instance, a uniform flexible string of length l, under tension and with fixed ends, has possible transverse vibrations of period $2\pi/\omega$ and displacement proportional to $\sin(\omega x/c)\cos\omega t$ (satisfying the one-dimensional wave equation with wave-speed c) *provided* $\omega l/c = k\pi$, where k is an integer (to satisfy the end-conditions). The *kth mode* thus has period $2l/kc$ and displacement proportional to $\sin(k\pi x/l)\cos(k\pi ct/l)$. Here the angular frequency $\omega_k = k\pi c/l$ is the *eigenvalue*, and the function $\sin(\omega_k x/c)$ the corresponding eigenfunction. $k = 1$ gives the *fundamental* – the mode of vibration with deepest note, longest period – and values of k greater than unity *overtones* or *harmonics*.

So far we have not introduced any initial conditions of motion. By suitable choice of such conditions (e.g. displacement $\sin(k\pi x/l)$ and velocity zero when $t = 0$) we can excite a single mode of vibration. But in general an initial displacement or impulse will excite many modes, and the subsequent vibration will be a superposition of those modes: if we listen to the vibration we will hear the fundamental and overtones.

The displacement $\sin(k\pi x/l)\cos(k\pi ct/l)$ could have been re-cast as

$$\frac{1}{2}\left\{\sin\frac{k\pi}{l}(x-ct)+\sin\frac{k\pi}{l}(x+ct)\right\}, \tag{1.2.1}$$

which we recognise as the superposition of two equal waves travelling in opposite directions with speed c. These two *travelling waves* cancel at $x=0$ and $x=l$, for all t, to form the *standing wave* $\sin(k\pi x/l)$ $\cos(k\pi ct/l)$. We will find the same duality in waves travelling through or around the Earth.

If the string has density varying smoothly along it, the differential equation of motion is more complicated, and the wave-speed varies with position, but the solution has essentially the same character as for a uniform string. The eigenfrequencies ω_k, when graphed against k, lie near a smooth curve, and ω_k/k tends to a certain limit as k tends to infinity. But if the density varies abruptly, as for instance in a string composed of two different halves, the pattern of eigenfrequencies changes. There is no longer a single asymptote, but a pattern slowly changing and indefinitely repeated. This is due to the existence of waves internally reflected at the discontinuity of the material constant. This observation is highly relevant to our study of vibrations of the Earth, since the Earth has internal regions where material constants change sharply enough to reflect elastic waves of suitable wave-lengths (see Chapter 8).

So far we have taken as example a system in which there is no loss of energy from internal friction. But all natural vibrating systems suffer attenuation; when the equation of motion is modified by the addition of a damping term, the solution for standing waves on a string with fixed end-points takes the form $e^{-\gamma t}\cos\omega' t\sin(k\pi x/l)$, where ω' is the modified angular frequency and γ the attenuation constant, both being functions of k. The amplitude of the standing oscillation decays by a factor e^{-1} in time γ^{-1}. A solution for a travelling wave may be found in the form $e^{-\alpha x}\cos(\beta x-\omega t)$, where α and β depend on ω. The amplitude decays by a factor e^{-1} in travelling a distance α^{-1}. This is a simple instance of the general effect of internal friction on vibrations.

One-dimensional vibrations are too simple to show all the characteristics of a general vibrating system. Let us increase the complexity by moving to vibrations of a rectangular membrane of sides a, b in length, fixed at the edges. The two-dimensional wave equation will be satisfied by a transverse displacement proportional to $\sin(m\pi x/a)\sin(n\pi y/b)\cos\omega t$, provided $\omega^2/c^2 = \pi^2(m^2/a^2 + n^2/b^2)$, where c is the wave-speed. The edge conditions demand that m and n are integers. Thus there is a doubly infinite set of eigenvalues, corresponding to all possible pairs of positive integers (m, n). To each pair corresponds a displacement pattern which is the eigenfunction, or mode, and an angular frequency $\omega_{m,n} = \pi c(m^2/a^2 + n^2/b^2)^{1/2}$.

There are two ways in which this system may degenerate. If we abandon conditions on the edges, $y = 0$ and $y = b$, and seek solutions in which there is no variation with y, so that displacement configurations are cylindrical, with generators parallel to the axis of y, the double infinity of eigenvalues (and corresponding eigenfunctions) reduces to a single infinity. If, on the other hand, we make $a = b$, the difference between $\omega_{m,n}$ and $\omega_{n,m}$ disappears. If we had started from a square and then lengthened one side, we would have seen each eigenfrequency split as the symmetrical square gave place to the unsymmetrical rectangle. We will find a similar phenomenon with vibrating elastic spheres.

Since an oscillating elastic solid is in general described by a three-dimensional wave equation, we will expect that there is a triply infinite set of normal modes with corresponding characteristic frequencies. Symmetries will reduce the set of frequencies, and so will restrictions (for instance, to purely radial displacements in a uniform sphere).

1.3 Early history of the study of free oscillations of the Earth

Kelvin reached his conclusions about the Earth's rigidity and frequency of free oscillation by a very indirect method. His analysis of the straining of an elastic sphere solved the statical problem, but he did not attack the dynamical. That problem waited for the brilliant and classic paper of Horace Lamb, published in 1882.

Lamb set up the equations of motion for small vibrations of a uniform elastic sphere of density ρ and Lamé's elastic constants λ and μ. He sought a harmonic solution containing a factor $\exp i\omega t$, and obtained a general solution of the field equations in terms of spherical harmonics. The condition that the surface of the sphere was stress-free led to one of two possible frequency equations, corresponding to vibrations of different types. In what Lamb called 'Vibrations of the First Class' there was no change of shape and no dilatation in twisting motions. In his 'Vibrations of the Second Class' the curl of the displacement (i.e. the local rotation) had no radial component (Stoneley, 1961). These classes are now called 'toroidal' (or sometimes 'torsional') and 'spheroidal' (or sometimes 'poloidal') respectively; they are described in detail in Chapter 2.

Lamb followed Kelvin in working with rectangular Cartesian coordinates. But for the geometry of a sphere it is usually simpler to use spherical polar coordinates (r, ϑ, φ). Then for either toroidal or spheroidal vibrations we can introduce eigenfunctions of the type $R(r)\Theta(\vartheta)\Phi(\varphi)$, where, to satisfy the wave equation and to make the solution single-valued in φ, Φ is a trigonometric function of $m\varphi$, m being an integer – the *longitudinal* (or azimuthal) order. Similarly Θ is an Associated Legendre function $P_l^m(\cos\vartheta)$, where l is an integer – the *latitudinal* (or *angular* or *Legendre*) degree. $R(r)$ then satisfies a Sturm–Liouville equation; it must be finite at the centre of the sphere, and must be such that the stress on an element of the surface of the sphere is zero. This poses an eigenvalue problem for ω. The solutions ${}_n\omega_l^m$, where n is the (radial) overtone number, form a triple array of eigenvalues. The arrays will differ for toroidal and spheroidal oscillations. The toroidal modes are denoted by ${}_nT_l^m$ and the spheroidal by ${}_nS_l^m$. This treatment, which is entirely equivalent to Lamb's method, was first given by Chree (1889). It is now the standard attack, and will be used throughout this book.

Lamb's beautiful paper solved completely the problem of vibrations of a perfectly elastic uniform non-gravitating, non-rotating sphere, but that was only a first step towards calculating the eigenfrequencies of the Earth. He had not considered self-gravitation, non-uniformity, rotation, ellipticity of meridian section or anelasticity. The hundred years since he wrote have seen the slow step-by-step conquest of the formidable difficulties presented by the real Earth.

In 1898 Bromwich investigated the effect of self-gravitation, rather naturally simplifying his work by dealing only with an incompressible sphere. This useful step opened the way for Love, who published in 1911 his splendid and authoritative monograph 'Some Problems of Geodynamics', containing a chapter on 'Vibrations of a gravitating compressible planet'. Love discovered that, provided the value of the rigidity μ was not too high, the sphere was capable of two types of spheroidal vibration – a slower one, in which gravity supplied the main restoring force, and a faster one, in which elastic recovery dominated. He calculated that 'for a homogeneous sphere of the same size as the Earth, having a rigidity equal to that of steel and a Poisson's ratio equal to $\frac{1}{4}$, the period of the slowest vibration of the type in question (spheroidal) is almost exactly 60 min. We see that the period is diminished by gravity, but not so much as it would be if the substance were incompressible.'

The discovery of the eigenfrequencies of the Earth depends on setting up field equations, and boundary conditions at the centre and surface, and identifying allowed frequencies of vibration of that dynamical system. The discussion of the bodily tide deals with the same dynamical system, but demands the response to given disturbing forces. For both problems the solution of the field equations with boundary conditions is equivalent to the solution of a variational problem. In its simplest form the variational problem is that of minimising the total energy of the system. Stoneley in 1926 formulated the statical problem in order to find the bodily tide. From about 1960 on, many authors have extended Stoneley's work to the dynamical problem (see §§ 1.4 and 6.1).

Before the invention of electronic computers it was not feasible to proceed to discussion of the oscillation of a radially heterogeneous sphere. Moreover, the fact that efforts to make direct observations of these oscillations for the Earth had failed relegated the question to a marginal area of theoretical seismology.

First observations of free oscillations of the Earth There is a Chinese record of the invention of a seismoscope by Chang Heng in the second century A.D., but the first efficient seismometers were built for measuring waves from earthquakes near the end of the nineteenth century, in Japan (Dewey & Byerly, 1969). The earliest indication that an earthquake shakes not simply the local region but the whole Earth was the identification in 1889 by Paschwitz (1893) at Potsdam, on the record of a horizontal pendulum built to search for Earth tides, of a pulse from an earthquake (of magnitude 5.8) in Japan (Knott, 1908, Chapter 11).

In 1925 Shida (1925) designed an electromagnetic seismograph with pendulum period 180 s and galvanometer period 1200 s for the purpose of observing the Earth's free oscillations; the technology of that time, however, was incapable of achieving the precision and sensitivity that were demanded. It was Benioff, an instrument designer of genius, who first constructed a strain-meter (a quartz rod 25 m long and anchored to the Earth at one end) which was capable of recording free vibrations as long as one hour (and also the bodily tide). This he set in an old mine tunnel at Isabella, California.

Then it was necessary to wait for a big earthquake. For, just as a large bell produces only some overtones and not the fundamental note in response to a small tap, so the Earth will produce its deeper notes when a large amount of kinetic energy is given to it by an earthquake or an explosion. The Kamchatkan earthquake of 1952 supplied such a burst of energy, and Benioff (Benioff, Gutenberg & Richter, 1954) found evidence in the Isabella seismogram of a motion with period 57 min. He suggested that this belonged to a grave characteristic mode of the Earth's vibration. Immediately theoreticians turned their attention to new computations, and experimentalists constructed new instruments – strain-meters and gravimeters – of such sensitivity to long period oscillations that they would respond to the Earth's deepest notes.

1.4 First computations of eigenfrequencies for a realistic Earth-model

Lamb and Love had confined their attention to uniform spheres. The first step towards calculation of eigenfrequencies for a realistic model Earth was by Matumoto & Satô (1954), who assumed two layers. They found that, for a model with uniform liquid core and uniform mantle in which the shear-wave velocity was 6.5 km/s, the slowest toroidal period was 42.5 min.

In 1956 Jobert published frequencies obtained by appeal to Rayleigh's Principle (see § 6.1). The great advantage of this Principle is that a first-order error in a trial eigenfunction leads to a *second-order* error in the derived eigenvalue. Jobert first established the usefulness of the method by calculations for simple models where the eigenvalue could be otherwise determined. She then obtained the fundamental period for a toroidal vibration of Legendre degree 2 for a Bullen model B Earth: it was 43.54 min. In 1957 she obtained, from the same method applied to the corresponding spheroidal mode, periods of about 53 min, the exact value depending on the value of the rigidity μ assumed for the Earth's inner core.

In 1959 Takeuchi was able to show how periods decrease with increasing Legendre degree. He used a Rayleigh–Ritz method and obtained for various Legendre degrees the following periods of toroidal oscillation:

Legendre degree (l)	2	4	8	16	32
Period (min)	43.43	21.48	12.11	7.02	3.96

At about the same time Pekeris, Alterman and colleagues embarked on the systematic attack on the eigenvalue computations, by numerical solution of differential equations, which is described in detail in Chapter 6. For toroidal oscillations they reduced the field equations to two simultaneous first-order ordinary differential equations. For spheroidal oscillations of a compressible self-gravitating Earth-model, Love's field equations were reduced to six simultaneous first-order ordinary differential equations (Alterman, Jarosch & Pekeris, 1959). The following periods were obtained for a Bullen B model, for $m = 0$ and four values of l in spheroidal oscillations:

Legendre degree l	0	2	3	4
Period of fundamental (min)	20.0	53.7	35.5	25.7
Period of first overtone (min)	10.0	24.7	17.9	14.4

The development of theory and elaboration of computations during the subsequent twenty years is described in later chapters.

1.5 The great Chilean and Alaskan earthquakes and the splitting of eigenfrequencies

By the date of the huge and devastating Chilean earthquake (22 May 1960) several observatories were equipped with strainmeters capable of detecting many of the longer free oscillations of the Earth. At the Helsinki meeting of the International Association of Seismology and Physics of the Earth's Interior (IASPEI) in July 1960, power spectra were exhibited which showed all the toroidal and spheroidal fundamentals up to ${}_0T_7^0$ and ${}_0S_7^0$ (see definitions in § 1.3),

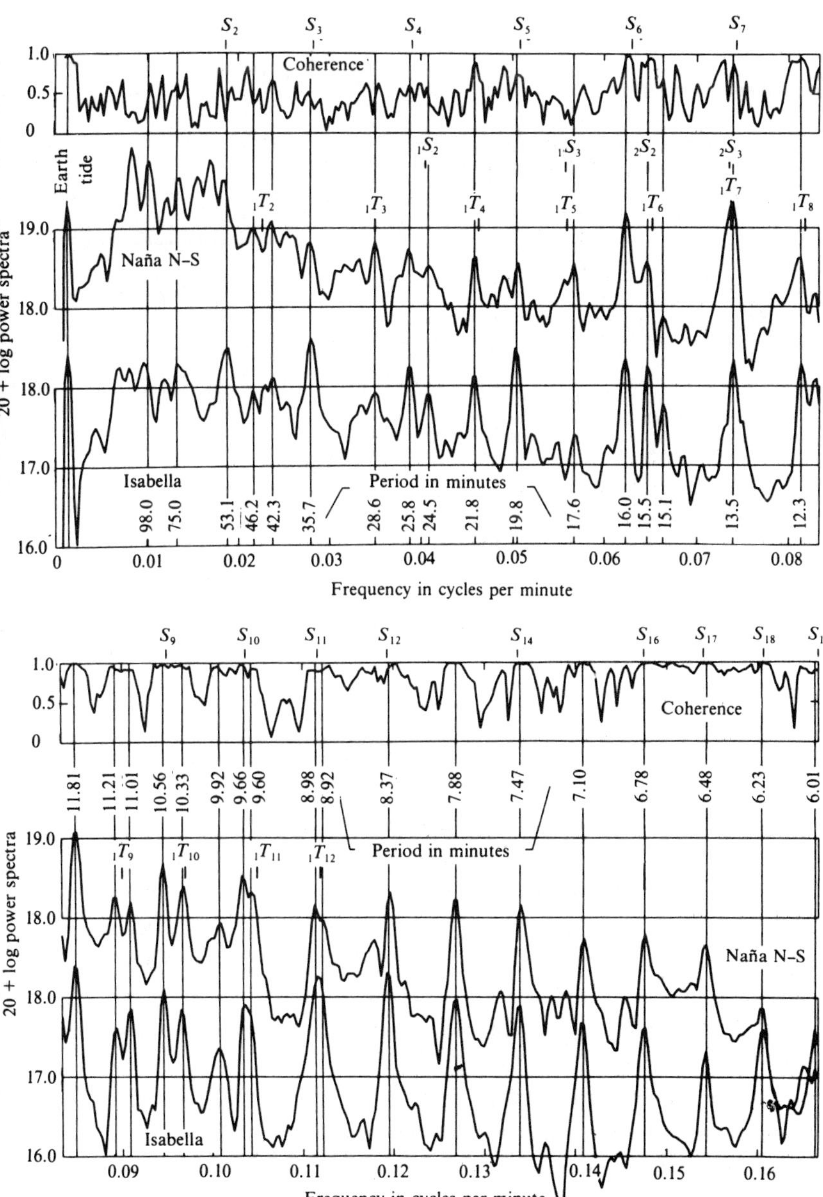

Fig. 1.1. Power spectra and coherence derived from strain seismograms of the Chilean earthquake (22 May 1960) at Isabella (California) and Naña (Peru). Spectral peaks corresponding to toroidal (T) and spheroidal (S) eigenfrequencies are identified. (After Benioff, Press & Smith, 1961.)

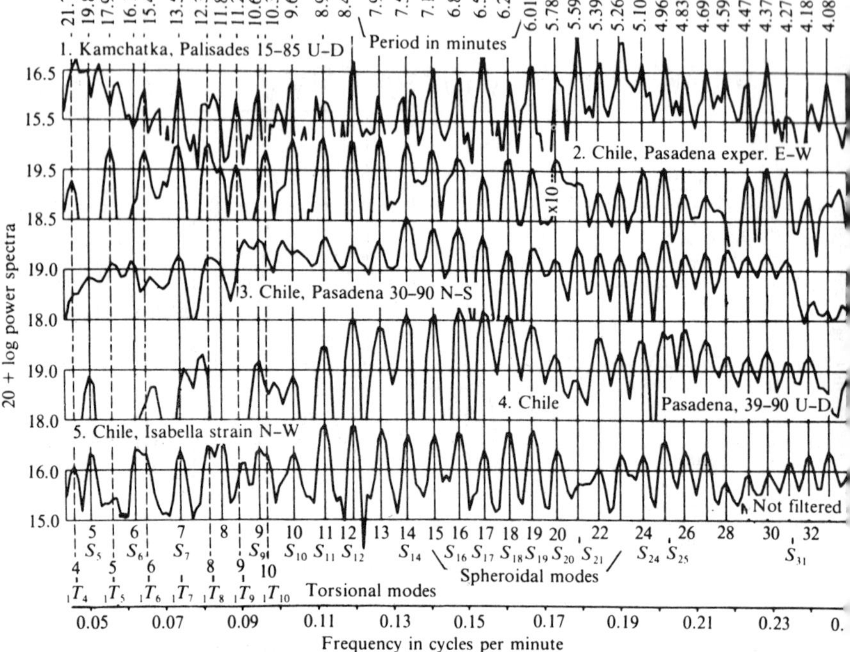

Fig. 1.2. Five power spectra, showing higher toroidal and spheroidal eigenfrequencies for Kamchatkan and Chilean earthquakes, recorded at Pasadena, Isabella and Palisades. (After Benioff *et al.*, 1961.)

identified after the Chilean earthquake. Slichter and colleagues at Los Angeles had found that their new gravimeter also had recorded some spheroidal oscillations. Changes in gravity do not accompany toroidal oscillations, and conversely a gravimeter should not register frequencies corresponding to toroidal oscillations, which were indeed absent from the spectrum. Thus strong evidence was provided that the identification of modes was correct (Ness, Harrison & Slichter, 1961).

Subsequent comparison of spectra from the Chilean earthquake obtained in Peru and California showed remarkable agreement (Fig. 1.1). Even more striking was the coincidence of spectral peaks in records from the Kamchatkan and Chilean earthquakes at California and New York (Fig. 1.2).

In 1964 the Alaskan earthquake occurred, of magnitude 8.4. By that time there were many instruments in operation competent to observe the Earth's eigenfrequencies. But the data suffered from the facts that the spectral peaks were broad and that no estimates of precision were available.

Great interest was aroused by the discovery that some of the peaks in the power spectra of the Chilean earthquake were multiple. This reminded physicists of the splitting of peaks in the well-known Zeeman effect, where atomic spectral lines are split when the emitting atoms lie in a magnetic field. Larmor (1897) had pointed out that the Zeeman effect is similar to that of a Coriolis force on a mechanical vibrator. The splitting of frequencies had been described for vibrations of a rotating liquid ellipsoid (Bryan, 1889) and of a rotating cylindrical bowl of water (Lamb, 1932). Quantitative explanations of the splitting of frequencies of the Earth's free oscillations were published by Pekeris, Alterman & Jarosch (1961a) and by Backus & Gilbert (1961).

In their analysis it became clear that the frequency equation could be perturbed not only by rotation but also by ellipticity of the Earth's meridian section. In fact the assumption of a non-rotating radially symmetric Earth-model produces a degenerate system in which the eigenfrequency is independent of m. But in the non-degenerate system to each value of l correspond the values $0, \pm 1, \pm 2, \ldots, \pm l$ of m. It was shown that each frequency spike calculated for the degenerate model should correspond to a set of $2l+1$ spikes in a rotating model, but only to $l+1$ in a non-rotating elliptic model.

Consider, as an analogy, a small ball rolling in a shallow dish which has the form of a horizontal spherical cap. The period of oscillation of the ball along the curve in which any vertical plane through the centre cuts the cap is the same: there is a single eigenperiod. But now let the dish be rotated uniformly about a vertical axis through the centre. It can easily be shown that instead of one characteristic period of the ball we now have two – one motion speeded up and one slowed down by the Coriolis force. Next revert to the stationary dish, but let horizontal sections be similar ellipses. There are again two normal modes with characteristic periods – the longer in the plane through major axes and the shorter in the plane through minor axes.

Similarly, any mode of vibration of a non-rotating axisymmetric sphere can have its degeneracy removed and frequency split by rotation or ellipticity. This question receives detailed treatment in Chapter 7.

1.6 Theoretical seismograms

Consider again the vibration of a flexible string of length l with fixed end-points. If we solve the field equation by separation of variables, and then force the solution to fit the end-conditions, we obtain, as a consistency condition, the frequency equation, the roots of which supply the eigenvalues. From these an infinite set of eigenfunctions may be defined. These describe all possible normal modes of vibration (i.e. modes in each of which every point of the string moves with the same frequency and phase).

But if in addition to the boundary conditions we prescribe *initial conditions* – for instance, by stating that the string is struck at a certain point at time $t = 0$ – then we must seek a linear combination of eigenfunctions which will satisfy the initial conditions. This linear combination of eigenfunctions will then describe exactly the configuration of the string at all subsequent times. Thus the solution of the eigenvalue problem is a first step to the solution of the *excitation problem*, that is, the problem of finding the response of the system to a stimulus given as function of position and time.

In exact analogy we can proceed beyond the discussion of eigenfrequencies of an Earth-model to the solution of the excitation problem, that is, to the construction of a linear combination of eigenfunctions which shall represent the response of the model to a stimulus which is a given function of position and time. The stimulus may be an explosion or an earthquake; at the simplest it may be a point impulse which is a spike function of time, but it may be a much more complicated simulation of an earthquake – for instance, a set of dislocations moving over a dipping plane of given orientation. The excitation problem is the subject of Chapter 9.

This may be called the *direct problem* of seismology. It has a unique solution provided the initial and boundary conditions are correctly specified. In essence it depends on our ability to find the response of radially heterogeneous sphere to a multipole or dislocation of arbitrary position and orientation and depending on time harmonically or as a delta-function. The complete solution for an extended fault can then be obtained by superposition (integration over space variables) and convolution with an appropriate function of time. When this has been done we obtain a *theoretical seismogram* or *synthetic seismogram* (Usami & Satô, 1972).

This is not the only way to construct a theoretical seismogram. Returning to the excitation of a string, we recall that instead of starting from the determination of eigenvalues and eigenfunctions we may utilise d'Alembert's solution of the wave equation

$$y = f(x - ct) + g(x + ct), \tag{1.6.1}$$

which represents pulses travelling in opposite directions with speed c. The forms of f and g are determined by the boundary and initial conditions. In this way we may construct the response to a unit source, which may be a displacement or an impulse, at an arbitrary point; this is *Green's function.* The response to a general initial disturbance may then be constructed by superposition of Green's functions.

The relation between these two methods has been one of the great seminal questions of applied mathematics. For simple problems we can show how to express Green's function as an infinite series of eigenfunctions, or to express a normal mode as a series of Green's functions. If in elastic wave theory we derive short-wave approximations to Green's functions, we find that these are the expressions for direct, reflected, refracted and diffracted rays which could have been constructed by geometrical-optical methods. The relations between rays and modes are discussed in Chapters 8 and 9.

1.7 The inverse problems of seismology

In solving direct problems of excitation, we start by assuming a model of the Earth's structure. But the most important objective of seismologists is to discover the (previously unknown) internal structure, starting from observations made at the surface of the Earth. This is the *inverse problem*, and one which far exceeds the direct problem in difficulty.

We have available as data the seismograms recorded at more

than a thousand stations scattered over the Earth's surface. These can be analysed to give times of arrival and amplitudes of seismic pulses for both body waves and surface waves – or else to give power spectra of free oscillations – set up by a seismic event. From these data we wish to find the distributions of density and elastic constants with depth within the Earth. For the time being we neglect complications due to transverse heterogeneity, ellipticity, rotation, anelasticity and anisotropy.

Since the data are incomplete – gathered at scattered points strongly biased towards positions on continents – it is clear that we cannot expect to determine uniquely a set of piecewise continuous parameters. But it is also clear that the observed data do greatly restrict the possible values of the parameters. How can we get the maximum amount of information about the internal structure of the Earth, bearing in mind that the data are subject to errors of observation?

The usual method is to take a trial model Earth as starting-point, and from this to compute (as in direct problems) quantities such as eigenfrequencies and travel-times. These can be compared with observed values, and the difference summarised (e.g. by the sum of squares of deviations). The trial model is then refined step by step to minimise the discrepancy. We may think of the problem as one of minimising an n-dimensional function, that is, of finding the lowest point of a surface in $(n+1)$ dimensions. There are various techniques for doing this numerically, but all depend on making a good first guess, so that the process is taking place in the true neighbourhood of the correct solution (compare Newton's method of finding approximately a root of an equation).

Each starting (trial) model of the Earth has certain explicit and implicit limitations, and it is found that refined models depend strongly on those limitations of the trial model, and also on the chosen measures of discrepancy. Thus it is hard to find an objective method of comparing models proposed by different workers. This topic is discussed in Chapter 10.

1.8 The solotone effect

Anderssen & Cleary (1974) examined the distribution of eigenfrequencies of radial overtones in toroidal oscillations of Earth-models. They pointed out that according to Sturm–Liouville theory this distribution should approach asymptotically, for large overtone number n, the value $n\pi/\gamma$, where γ is the time taken by a shear wave to travel along a radius from the core–mantle interface to the surface, *provided* elastic parameters and their derivatives vary continuously along the radius (Fix, 1967). They found that, for all the models which they considered, the distribution of eigenfrequencies deviated from the asymptote by amounts which seemed to depend on the existence and size of internal discontinuities. When Sturm–Liouville theory was extended to deal with systems containing internal discontinuities (McNabb, Anderssen & Lapwood, 1976), it was shown that the pattern of deviations – the *solotone effect* – could be predicted from knowledge of the position and size of the discontinuities.

It can be shown that the solotone effect, exhibited by the graph of $(\gamma_n \omega_l / \pi) - n$ (Fig. 8.3), oscillates with amplitudes determined by the magnitude of the discontinuities, but periods and phases determined by the thicknesses and wave-speeds of the layers. It appears that the pattern of eigenfrequencies is very sensitive to small changes in relative thickness of the layers. Further examination reveals how each component sinusoidal oscillation in the solotone effect is the result of resonance between waves which, having taken different paths, interfere constructively.

The identification of peaks in power spectra due to higher radial overtones is difficult, and as yet not far advanced. But, given any proposed Earth-model with prescribed surfaces of discontinuity, it is possible to predict the corresponding solotone pattern. In due course this may form one of the ways of discriminating between models.

This account has dealt with toroidal oscillations. It is to be expected that spheroidal oscillations will show similar effects, with additional complications. Solotone effects are described in Chapter 8.

1.9 Current work

So far we have been discussing properties of Earth-models with axial symmetry. That the Earth itself departs significantly from such symmetry at the crust can be seen from the distribution of oceans and continents and from the large variations in depth of the Mohorovičić discontinuity. That it departs from symmetry at depth is shown by the pattern of deep earthquake foci and by the differences in travel-time tables for P in different regions (Jeffreys, 1976, § 3.101). These deviations from axial symmetry will cause the response to a given seismic event to depend on the positions of focus and point of observation; the dependence will show itself, for short waves, in irregular travel-times, and for long waves and free oscillations in variations of spectral amplitudes.

The study of regional variations in travel-times has a long history: the classical method of approach has been to select suitable events and observing stations in a region assumed, on geological evidence, to be fairly uniform in structure, and to fit travel-time curves for P and S. These are then inverted to provide tables of velocities as functions of depth for that region. Other methods recently employed are described in Chapter 10, along with modern studies of inhomogeneity in longitude in relation to free oscillations.

A second possible source of discrepancy between our models and the real Earth lies in our assumption that even though the Earth is not uniform it is isotropic. One would expect this assumption to be false if the material of the mantle was ever molten, since convective currents and magnetic fields would introduce directional differences in the physical constants of the rock at the time of solidification. Some evidence has been put forward (Crampin, 1977) that the upper mantle is anisotropic in at least some regions.

A third departure of our models from the real Earth starts from our hypothesis of perfect elasticity. From the fact that all the vibrations with which we have been dealing are seen to be attenuated we know that the material of the Earth is *anelastic*. The surprising thing is not that mechanical energy is dissipated during vibration, but that seismologists have been able to get so far in matching theory to observations while using classical analysis for a perfectly elastic medium.

It is not easy to decide what physical mechanism is responsible for the decay of mechanical energy: there may be more than one. But fortunately a fairly general hypothesis of linear viscoelasticity enables us to allow for anelasticity, at least where it is small, by taking the elastic moduli and consequently the velocities as complex numbers depending on frequency. We can then prove that any travelling wave which is attenuated in amplitude must also suffer change in phase. In a wave packet composed of different frequencies the amplitude and wave-speed of each component will suffer frequency-dependent change and the packet will be dispersed. The frequencies of modes of standing waves will be reduced by anelasticity.

Jeffreys has maintained for many years (Jeffreys, 1967) that this fact makes reliance on observations of free oscillation periods wrong, if no allowance is made for anelasticity, in construction of Earth-models. Recent work by Randall (1976) and Kanamori & Anderson (1977) has confirmed Jeffreys's objection, and has shown that the discrepancies between models based on free oscillation data and models based on travel-time data can be removed, and quantitative assessment of anelasticity made. Anelastic behaviour of a vibrating medium may be characterised by a *quality factor* Q, where Q^{-1} measures the proportionate loss of energy in one cycle. Materials with high Q ring for a long time, those with low Q experience rapid damping. The distribution of Q with depth in the mantle can be estimated by watching the attenuation with time of normal modes of free oscillation. As will be shown later, different modes concentrate energy at different depths, so that Qs referring to those depths are found.

During the past 30 years many different Earth-models have been proposed. They have been very difficult to compare, on account of differences in assumptions, methods, and styles of presentation. It was proposed by Bullen and others that international agreement should be sought for the designation of a single reference model, to which future proposed Earth-models could be related. The Standard Earth Model Committee was set up and assigned this task by the International Union of Geodesy and Geophysics in 1974. Its work, still unfinished, is described in Chapter 10.

2

FREE OSCILLATION OF A UNIFORM ELASTIC SPHERE

2.1 Introductory remarks

In this chapter we will study in some detail the free oscillation of a homogeneous and isotropic (uniform) elastic sphere. Using this simplest model we will describe the fundamental properties necessary for an understanding of the physical meaning of the free oscillations. We assume that readers know the equation of motion of an elastic medium.

The first step in the study of free oscillations is to calculate the eigenfrequency and eigenfunction of each mode: this can be done by solving the equation of motion under suitable boundary conditions. We find that there are two kinds of oscillation, and that each mode of either kind can be identified by the three suffixes l, m and n ($l, n = 0, 1, 2, \ldots ; |m| = 0, 1, \ldots, l$). We shall describe general properties of each mode as functions of l, m and n, and explain the meaning of these three suffixes in relation to eigenfunctions and surface waves. It will be shown that each mode of free oscillation can be considered as a pair of surface waves travelling along a meridian on the surface of the sphere, and the relation between free oscillation and surface waves will be explored in detail.

It is well-known that there are two methods of solving a problem of wave propagation, that is, in terms of travelling waves or of standing waves (modes). If the computation is carried out completely, both theories should give the same results. However, in travelling-wave theory (and in ray-theory which starts from the short-wave approximation) contributions of suitable waves are summed to give the disturbance, whilst in mode-theory contributions from all modes are summed. Here the question arises: what is the exact correspondence between modes and rays? In spite of the importance of this problem, that correspondence is not yet completely clear. In this chapter we will study briefly some basic correspondence relations between the two theories.

The results found in this chapter for a uniform sphere are fundamental for the understanding of the physical aspects of free oscillation of all spheres, and can be applied to more complicated models with slight modifications.

2.2 Equation of motion of a uniform elastic medium and its solution

The equations of small motion of a uniform perfectly elastic medium, in Cartesian coordinates, are (Jeffreys, 1976, § 1.02)

$$\partial_j \tau_{ij} + F_i = \rho \partial_t^2 u_i, \tag{2.2.1}$$

where τ_{ij} is the stress tensor, F_i the body force per unit volume, ρ the density, u_i the displacement and x_j (implicitly involved in the notation $\partial_j \equiv \partial/\partial x_j$) the position vector. The usual summation convention of tensor analysis is assumed. The first term of the left-hand side of (2.2.1) is a force in the x_i direction due to stress gradient and the right-hand side is the inertial term – mass-acceleration. Introducing the stress–strain relation, that is, the generalisation of Hooke's law for a perfectly elastic isotropic material,

$$\tau_{ij} = \lambda \Delta \delta_{ij} + 2\mu e_{ij}, \tag{2.2.2}$$

we write (2.2.1) as

$$\partial_j(\lambda \Delta \delta_{ij} + 2\mu e_{ij}) + F_i = \rho \partial_t^2 u_i, \tag{2.2.3}$$

where λ and μ are Lamé's constants, Δ is the dilatation

$$\Delta = \partial_k u_k, \tag{2.2.4}$$

and e_{ij} the strain tensor

$$e_{ij} = \tfrac{1}{2}(\partial_j u_i + \partial_i u_j). \tag{2.2.5}$$

If the medium is *uniform*, ρ, λ and μ are constant throughout the medium, and if the medium is free from body force, (2.2.3) reduces to

$$(\lambda + 2\mu)\,\mathrm{grad}\,\mathrm{div}\,\mathbf{u} - \mu\,\mathrm{curl}\,\mathrm{curl}\,\mathbf{u} = \rho \partial_t^2 \mathbf{u}, \tag{2.2.6}$$

where $\mathbf{u}$ is the displacement vector.

Any vector $\mathbf{K}$ may be expressed, by Helmholtz's theorem (Morse & Feshbach, 1953, § 1.5), as the sum of a gradient of a scalar ϕ and a curl of a vector $\mathbf{B}$,

$$\mathbf{K} = \mathrm{grad}\,\phi + \mathrm{curl}\,\mathbf{B}. \tag{2.2.7}$$

Three scalar functions are necessary to specify a vector in three-dimensional space. One of the three scalar functions is the scalar potential ϕ, so that only two scalar functions should specify the vector curl **B**. This gives some freedom to **B**, which we usually use to make div $\mathbf{B} = 0$.

With this theorem in mind, let us try to obtain a solution of (2.2.6) in the form

$$\mathbf{u} = \operatorname{grad} \phi + \operatorname{curl} \mathbf{B}. \tag{2.2.8}$$

We begin with the solution where $\mathbf{B} = 0$:

$$\mathbf{u}_1 = \operatorname{grad} \phi. \tag{2.2.9}$$

We look for a free oscillation of angular frequency ω, so that $\partial_t^2 \equiv -\omega^2$. Then from (2.2.6) we have

$$(\lambda + 2\mu) \operatorname{grad}(\operatorname{div} \operatorname{grad} \phi) + \rho\omega^2 \operatorname{grad} \phi = 0. \tag{2.2.10}$$

If we introduce the notation ∇^2 (Appendix § A.1):

$$\nabla^2 \phi \equiv \operatorname{div} \operatorname{grad} \phi, \tag{2.2.11}$$

it is sufficient that

$$\nabla^2 \phi + h^2 \phi = 0, \tag{2.2.12}$$

where

$$h = \omega / c_{\mathrm{P}}, \tag{2.2.13}$$

and

$$c_{\mathrm{P}} = \{(\lambda + 2\mu)/\rho\}^{1/2} \tag{2.2.14}$$

is the speed of the (longitudinal) P wave.

In order to construct the vector potential in (2.2.8) let us consider the boundary conditions. One possible set of boundary conditions would be that the normal and tangential components of stress on the boundary surface should vanish. These are in fact the boundary conditions required for free oscillations. If the first term in (2.2.8), grad ϕ, is called the *normal* component (since the gradient points in the direction of greatest rate of change of the scalar potential ϕ), then the second term may be called the *transverse* component (since the curl of a vector is usually transverse to the direction of greatest change).† For plane waves these components are in and orthogonal to the direction of propagation. (For details, refer to Morse & Feshbach (1953), §§ 13 and 13.1.)

Let us therefore try, as a second component, a vector

$$\mathbf{u}_2 = \operatorname{curl} \mathbf{B}, \quad \mathbf{B} = (r, 0, 0)_{\text{polar}} \psi, \tag{2.2.15}$$ ††

where ψ is a scalar function of position. This vector $\mathbf{u}_2$ is tangential to the spherical surface $r = \text{const}$, and it will be a solution if ψ is such as to make $\mathbf{u}_2$ satisfy the equation of motion (2.2.6). Putting (2.2.15) with (2.2.6), it is found to be sufficient, for $\mathbf{u}_2$ to be a solution, that

$$\nabla^2 \psi + k^2 \psi = 0, \tag{2.2.16}$$

where $k = \omega / c_S$ is the wave-number and

$$c_S = (\mu/\rho)^{1/2} \tag{2.2.17}$$

is the speed of an S (shear) wave. In the reduction of (2.2.16), formulae (A1.2–4) in Appendix § A.1 are employed. This is one of the two scalar fields required to generate a transverse solution. One more solution is needed. It should, if possible, generate a field orthogonal to $\mathbf{u}_1$ and $\mathbf{u}_2$. Let us try a vector resembling the curl of $\mathbf{u}_2$, namely

$$\mathbf{u}_3 = \frac{1}{k} \operatorname{curl} \operatorname{curl} \mathbf{C}, \qquad \mathbf{C} = (r, 0, 0)_{\text{polar}} \chi, \tag{2.2.18}$$

where χ is a scalar function of position. The curl of $\mathbf{C}$ is tangential to spherical surfaces $r = \text{const}$, and

$$\nabla^2 \chi + k^2 \chi = 0 \tag{2.2.19}$$

is sufficient for $\mathbf{u}_3$ to satisfy the equation of motion (2.2.6).

We have thus defined three scalar fields ϕ, ψ and χ, all solutions of the scalar Helmholtz equation. A general solution $\mathbf{u}$ of (2.2.6) is then expressed as

$$\mathbf{u} = \mathbf{u}_1 + \mathbf{u}_2 + \mathbf{u}_3,$$

†Let χ be a scalar function and $\hat{\mathbf{e}}$ be an unit vector in a certain direction, then we have curl $\chi\hat{\mathbf{e}} = \operatorname{grad} \chi \wedge \hat{\mathbf{e}}$.

††In (2.2.15) and (2.2.18) we have introduced vectors in the direction of the radius to obtain **B** and **C**. Other choices are possible, as shown by Usami (1962) and Usami, Kano & Satô (1962). In any particular problem the type of boundary conditions determines the appropriate choice of vectors.

where

$$\mathbf{u}_1 = \text{grad}\,\phi, \tag{2.2.20}$$

$$\mathbf{u}_2 = \text{curl}(r\psi, 0, 0)_{\text{polar}} \tag{2.2.21}$$

$$\mathbf{u}_3 = \frac{1}{k}\,\text{curl curl}(r\chi, 0, 0)_{\text{polar}} \tag{2.2.22}$$

The factor $1/k$ is made explicit in $\mathbf{u}_3$, so that the dimensions of $\phi, r\psi$ and $r\chi$ are all the same. It is clear that $\text{curl}\,\mathbf{u}_1 = 0$ and $\text{div}\,\mathbf{u}_2 = 0$, $\text{div}\,\mathbf{u}_3 = 0$. Thus, $\mathbf{u}_1$ represents an irrotational wave and $\mathbf{u}_2$ and $\mathbf{u}_3$ equivoluminal waves. Moreover, the radial component of curl $\mathbf{u}_3$ vanishes.

If the Helmholtz equations (2.2.12), (2.2.16) and (2.2.19) are solved by the method of separation of variables (Morse & Feshbach, 1953, § 11.3, p. 1462), we obtain

$$\phi = z_l(hr)P_l^m(\cos\vartheta)\,{}^{\cos}_{\sin}\,m\varphi\,\mathrm{e}^{\mathrm{i}\omega t}, \tag{2.2.23}$$

where $z_l(hr)$ is a spherical Bessel function which satisfies the differential equation

$$\partial_r^2 z_l + \frac{2}{r}\partial_r z_l + \left\{h^2 - \frac{l(l+1)}{r^2}\right\} z_l = 0. \tag{2.2.24}$$

For potentials ψ and χ, k replaces h. $P_l^m(\cos\vartheta)$ is the Associated Legendre function

$$P_l^m(\cos\vartheta) = \sin^m\vartheta\left(\frac{\mathrm{d}}{\mathrm{d}\cos\vartheta}\right)^m P_l(\cos\vartheta) \qquad (m \le l) \tag{2.2.25}$$

(Ferrers' definition), which satisfies the differential equation

$$\partial_\vartheta^2 P_l^m + \cot\vartheta\,\partial_\vartheta P_l^m + \left\{l(l+1) - \frac{m^2}{\sin^2\vartheta}\right\} P_l^m = 0. \tag{2.2.26}$$

The argument $\cos\vartheta$ will be suppressed as long as there is no confusion.

The trigonometric functions

$$\cos m\varphi, \qquad \sin m\varphi, \qquad m = 0, 1, 2, 3, \ldots$$

form a complete set of orthogonal eigenfunctions of the differential equation

$$\mathrm{d}^2\Phi/\mathrm{d}\varphi^2 + m^2\Phi = 0.$$

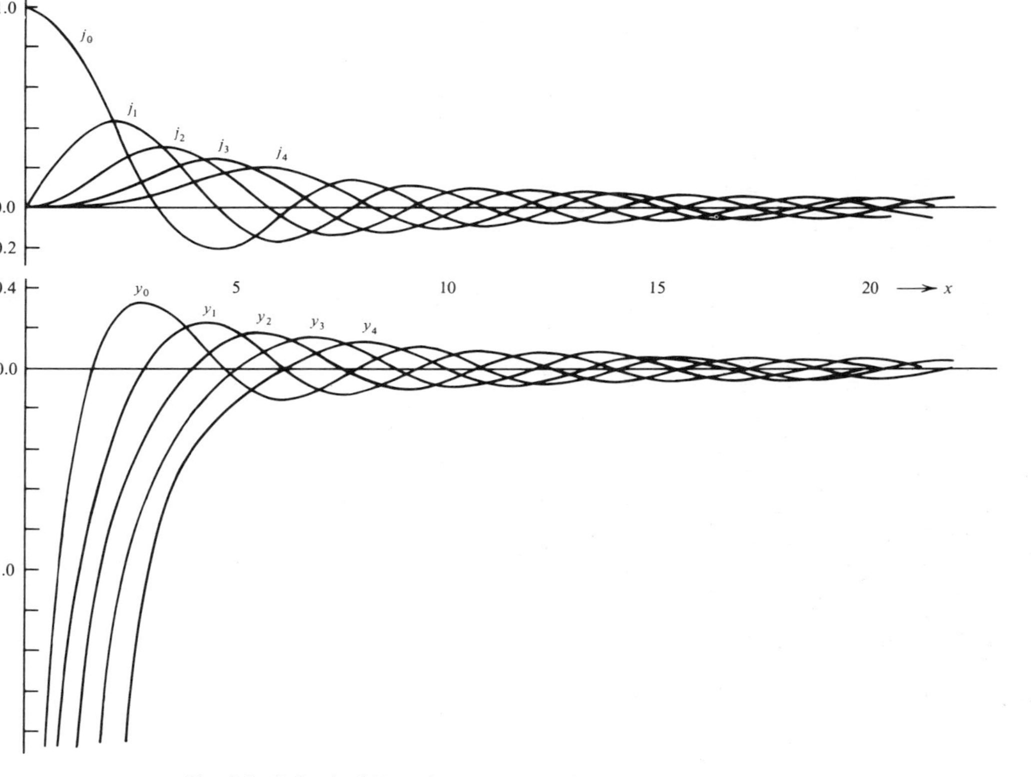

Fig. 2.1. Spherical Bessel functions j_l and y_l for $l = 0, 1, 2, 3, 4$.

The Associated Legendre functions also form a complete orthogonal set of eigenfunctions of the equation (2.2.26) provided l and m are integers and $m \leq l$. Equation (2.2.24) is of Sturm–Liouville type; with the appropriate boundary conditions it determines an infinite set of eigenfrequencies ${}_n\omega_l$. The eigenvalues of h are then ${}_n\omega_l/c_{\mathrm{P}}$, and the corresponding eigenfunctions (functions of r) again form a complete orthogonal set. In each case the eigenfunctions can be normalised by introduction of an appropriate constant factor. Thus we have built up a set of eigenfunctions of the original wave equation, each defined by a triplet (l, m, n), in terms of which a given function of position can be expressed.

z_l takes the form j_l or y_l according to the physical conditions imposed. j_l and y_l are linearly independent for all l and are such that, for large x,

$$j_l(x) \simeq (1/x)\sin(x - \tfrac{1}{2}l\pi), \qquad y_l(x) \simeq -(1/x)\cos(x - \tfrac{1}{2}l\pi). \quad (2.2.27)$$

Fig. 2.1 shows the form of j_l and y_l for some values of l.

$P_l^m(\cos\vartheta)^{\cos}_{\sin} m\varphi$ are *spherical harmonics*, those for $m=0$ being *zonal harmonics* (Fig. 2.2(*a*)), those for $m=l$ being *sectorial harmonics* (Fig. 2.2(*b*)) and the rest being *tesseral harmonics* (Fig. 2.2(*c*), (*d*)). The combination of l and m determines the pattern of motion on the spherical surface determined by any value of r.

Full expressions for (2.2.20–22), which are used in the next section when we compute eigenfrequencies, are

$$\text{(irrotational)} \quad \left.\begin{aligned} u_{1r} &= u_1 = A_{ml}\frac{\mathrm{d}}{\mathrm{d}r}z_{1l}P_l^m {}^{\cos}_{\sin} m\varphi\, \mathrm{e}^{\mathrm{i}\omega t}, \\ u_{1\vartheta} &= v_1 = A_{ml}\frac{1}{r}z_{1l}\partial_\vartheta P_l^m {}^{\cos}_{\sin} m\varphi\, \mathrm{e}^{\mathrm{i}\omega t}, \\ u_{1\varphi} &= w_1 = -mA_{ml}\frac{1}{r}z_{1l}\frac{1}{\sin\vartheta} \\ &\quad \times P^m_l {}^{\sin}_{(-\cos)} m\varphi\, \mathrm{e}^{\mathrm{i}\omega t}, \end{aligned}\right\} \quad (2.2.28)$$

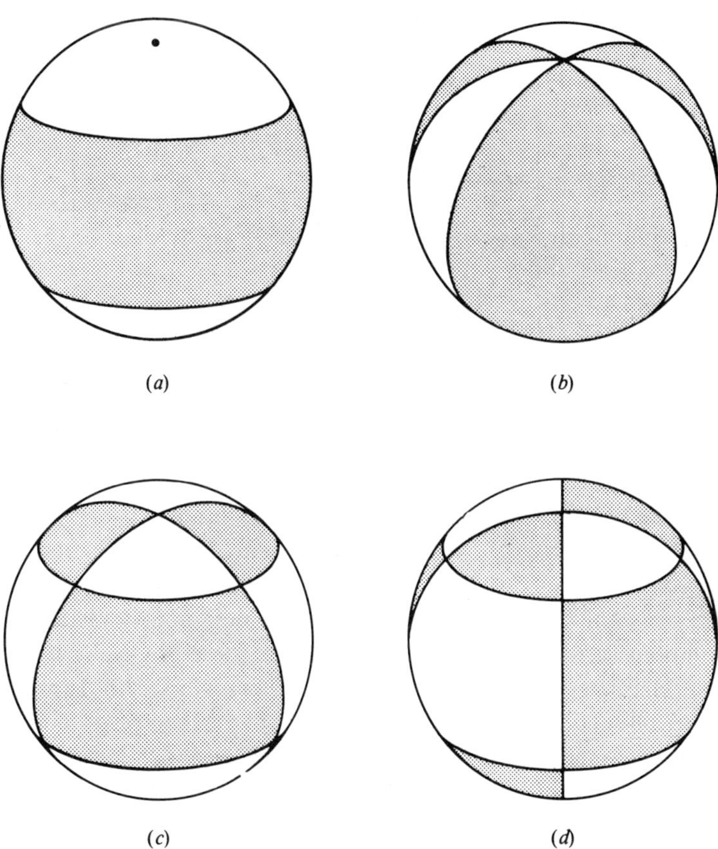

Fig. 2.2. Spherical harmonics. Hatched and blank regions show different signs.

(a) $P_2^0 = \frac{1}{4}(1 + 3\cos 2\vartheta)$,

(b) $P_3^3 \cos 3\varphi = 15 \sin^3 \vartheta \cos 3\varphi$,

(c) $P_4^2 \cos 2\varphi = \frac{15}{16}(3 + 4\cos 2\vartheta - 7\cos 4\vartheta)\cos 2\varphi$,

(d) $P_4^2 \sin 2\varphi = \frac{15}{16}(3 + 4\cos 2\vartheta - 7\cos 4\vartheta)\sin 2\varphi$.

$$\text{(equivoluminal)}\left\{\begin{aligned} u_{2r} &= u_2 = 0, \\ u_{2\vartheta} &= v_2 = mB_{ml} z_{2l} \frac{1}{\sin\vartheta} P_l^m \,{}^{\sin}_{(-\cos)} m\varphi\, \mathrm{e}^{\mathrm{i}\omega t}, \\ u_{2\varphi} &= w_2 = B_{ml} z_{2l} \partial_\vartheta P_l^m \,{}^{\cos}_{\sin} m\varphi\, \mathrm{e}^{\mathrm{i}\omega t} \end{aligned}\right\} \quad (2.2.29)$$

$$\text{(equivoluminal)}\left\{\begin{aligned} u_{3r} &= u_3 = C_{ml} \frac{l(l+1)}{kr} z_{3l} P_l^m \,{}^{\cos}_{\sin} m\varphi\, \mathrm{e}^{\mathrm{i}\omega t}, \\ u_{3\vartheta} &= v_3 = C_{ml} \frac{1}{kr} \frac{\mathrm{d}}{\mathrm{d}r}(r z_{3l}) \partial_\vartheta P_l^m \,{}^{\cos}_{\sin} m\varphi\, \mathrm{e}^{\mathrm{i}\omega t}, \\ u_{3\varphi} &= w_3 = -mC_{ml} \frac{1}{kr} \frac{\mathrm{d}}{\mathrm{d}r}(r z_{3l}) \\ &\quad \times \frac{1}{\sin\vartheta} P_l^m \,{}^{\sin}_{(-\cos)} m\varphi\, \mathrm{e}^{\mathrm{i}\omega t}, \end{aligned}\right\} \quad (2.2.30)$$

where z_{1l}, z_{2l}, z_{3l} stand for j_l or y_l according as we use them in a region containing the origin or extending to infinity. The argument in z_{1l} is hr and that in both z_{2l} and z_{3l} is kr. In the reduction, the differential equation (2.2.26) has been employed. A_{ml}, B_{ml} and C_{ml} are constant multipliers which are to be determined so that the solutions satisfy the boundary conditions. Some physically important characteristics of $\mathbf{u}_1$, $\mathbf{u}_2$ and $\mathbf{u}_3$ are summarised in Table 2.2 (§ 2.5). Functions of φ in (2.2.18), (2.2.30) show that, for $\mathbf{u}_1$ and $\mathbf{u}_3$, the argument of radial and latitudinal displacement component is in advance by $\frac{1}{2}\pi$ of that of the azimuthal, and for $\mathbf{u}_2$ the argument of the azimuthal displacement component is in advance by $\frac{1}{2}\pi$ of that of the latitudinal. These differences of argument and functions of ϑ are combined to give patterns of surface displacement distributions characteristic of different types of oscillation (see Figs. 2.5 and 2.6).

2.3 Free oscillations – eigenfrequency and eigenfunction

In Chapter 1 the concept of free oscillation was explained, with one- and two-dimensional examples. Similarly, in a three-dimensional medium we investigate free oscillations by solving the equation of motion under suitable boundary conditions. For a spherical elastic medium, the boundary conditions which we wish

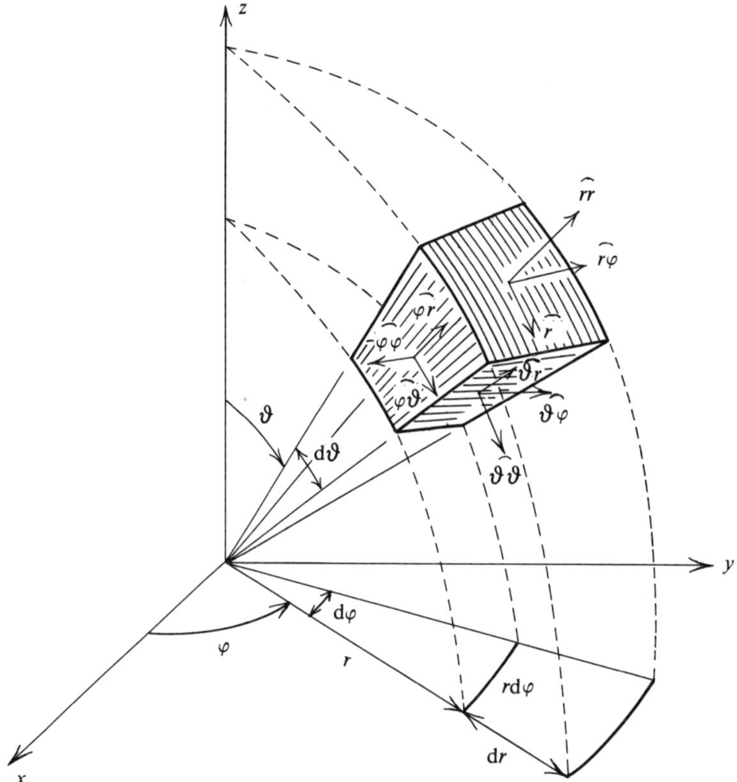

Fig. 2.3. Stress components in spherical polar coordinates.

to employ are the regularity of displacement components at the centre $r = 0$ and the continuity of stress components on the surface of the sphere. Since the stress components due to air pressure at the surface of the Earth are negligibly small, the latter conditions will be replaced by the vanishing of the stress components on the surface. Fig. 2.3 shows stress components referred to spherical polar coordinates. Thus we want

$$u, v, w \text{ finite at } r = 0, \tag{2.3.1}$$

and stress components

$$\widehat{rr} = 0, \quad \widehat{r\vartheta} = 0, \quad \widehat{r\varphi} = 0 \text{ at } r = a, \tag{2.3.2}$$

where a is the radius of the sphere, and $\widehat{rr}, \widehat{r\vartheta}, \ldots$ denote $\tau_{rr}, \tau_{r\vartheta}, \ldots$.

If we denote the components of the strain tensor by $e_{rr}, \dots$, and the dilatation by $\Delta = e_{rr} + e_{\vartheta\vartheta} + e_{\varphi\varphi}$, then, for a medium which is perfectly elastic and isotropic, the stress–strain relations for small strains take the form

$$\left.\begin{aligned} \widehat{rr} &= \lambda\Delta + 2\mu e_{rr}, \\ \widehat{\vartheta\vartheta} &= \lambda\Delta + 2\mu e_{\vartheta\vartheta}, \\ \widehat{\varphi\varphi} &= \lambda\Delta + 2\mu e_{\varphi\varphi}, \end{aligned}\right\} \tag{2.3.3}$$

$$\left.\begin{aligned} \widehat{r\vartheta} &= 2\mu e_{r\vartheta}, \\ \widehat{\vartheta\varphi} &= 2\mu e_{\vartheta\varphi}, \\ \widehat{\varphi r} &= 2\mu e_{\varphi r}, \end{aligned}\right\} \tag{2.3.4}$$

where λ and μ are Lamé's constants.

The components e_{ij} of the strain tensor are given (Appendix § A.2) by

$$\left.\begin{aligned} e_{rr} &= \partial_r u, \\ e_{\vartheta\vartheta} &= \frac{1}{r}\partial_\vartheta v + \frac{1}{r}u, \\ e_{\varphi\varphi} &= \frac{1}{r\sin\vartheta}\partial_\varphi w + \frac{1}{r}u + \frac{\cot\vartheta}{r}v, \\ e_{r\vartheta} &= \frac{1}{2}\left(\frac{1}{r}\partial_\vartheta u + \partial_r v - \frac{1}{r}v\right), \\ e_{\vartheta\varphi} &= \frac{1}{2}\left(\frac{1}{r\sin\vartheta}\partial_\varphi v + \frac{1}{r}\partial_\vartheta w - \frac{\cot\vartheta}{r}w\right), \\ e_{\varphi r} &= \frac{1}{2}\left(\frac{1}{r\sin\vartheta}\partial_\varphi u + \partial_r w - \frac{1}{r}w\right), \end{aligned}\right\} \tag{2.3.5}$$

where u, v, w are here used to express displacement components in r, ϑ and φ directions. Curvature of the coordinate lines introduces the terms which do not contain derivatives in (2.3.5). Some of the components set out in (2.3.3–5) are not necessary for the boundary conditions, but all components are needed in the equations of motion.

Full expressions for the stress components necessary for our study are as follows. For the solution $\mathbf{u}_1$, we have, using ∂_r or $(\dot{\ })$ for d/dr as convenient,

$$
\text{(irrotational)} \left\{
\begin{aligned}
\widehat{rr}_1 &= -A_{ml}\left\{(\lambda+2\mu)h^2 z_{1l} + \frac{4\mu}{r}\dot{z}_{1l} - \frac{2\mu l(l+1)}{r^2} z_{1l}\right\} P_l^m \tfrac{\cos}{\sin} m\varphi\, e^{i\omega t},\\
\widehat{r\vartheta}_1 &= 2\mu A_{ml}\,\partial_r\left(\frac{1}{r} z_{1l}\right)\partial_\vartheta P_l^m \tfrac{\cos}{\sin} m\varphi\, e^{i\omega t},\\
\widehat{r\varphi}_1 &= -2m\mu A_{ml}\,\partial_r\left(\frac{1}{r} z_{1l}\right) \times \frac{1}{\sin\vartheta} P_l^m \tfrac{\sin}{(-\cos)} m\varphi\, e^{i\omega t}.
\end{aligned}
\right\} \tag{2.3.6}
$$

For the solutions $\mathbf{u}_2$ and $\mathbf{u}_3$, we have

$$
\begin{array}{l}\text{(equivoluminal,}\\ \text{with zero radial}\\ \text{displacement)}\end{array} \left\{
\begin{aligned}
\widehat{rr}_2 &= 0,\\
\widehat{r\vartheta}_2 &= m\mu B_{ml}\left(\dot{z}_{2l} - \frac{1}{r} z_{2l}\right) \times \frac{1}{\sin\vartheta} P_l^m \tfrac{\sin}{(-\cos)} m\varphi\, e^{i\omega t},\\
\widehat{r\varphi}_2 &= \mu B_{ml}\left(\dot{z}_{2l} - \frac{1}{r} z_{2l}\right)\partial_\vartheta P_l^m \tfrac{\cos}{\sin} m\varphi\, e^{i\omega t},
\end{aligned}
\right\} \tag{2.3.7}
$$

$$
\text{(equivoluminal)} \left\{
\begin{aligned}
\widehat{rr}_3 &= 2\mu C_{ml}\frac{l(l+1)}{k}\partial_r\left(\frac{1}{r} z_{3l}\right) \times P_l^m \tfrac{\cos}{\sin} m\varphi\, e^{i\omega t},\\
\widehat{r\vartheta}_3 &= \mu C_{ml}\frac{1}{k}\left\{\ddot{z}_{3l} + \frac{l(l+1)-2}{r^2} z_{3l}\right\} \times \partial_\vartheta P_l^m \tfrac{\cos}{\sin} m\varphi\, e^{i\omega t},\\
\widehat{r\varphi}_3 &= -m\mu C_{ml}\frac{1}{k}\left\{\ddot{z}_{3l} + \frac{l(l+1)-2}{r^2} z_{3l}\right\} \times \frac{1}{\sin\vartheta} P_l^m \tfrac{\sin}{(-\cos)} m\varphi\, e^{i\omega t}.
\end{aligned}
\right\} \tag{2.3.8}
$$

As stated above, z_{1l} stands for $z_{1l}(hr)$ and z_{2l}, z_{3l} for $z_{2l}(kr)$, $z_{3l}(kr)$. $\widehat{rr}_2$ is zero because both u and $\operatorname{div}\mathbf{u}_2 = \Delta$ vanish for the solution $\mathbf{u}_2$.

Boundary conditions (2.3.2) at the free surface are now

$$\left.\begin{aligned} \widehat{rr}_1+\widehat{rr}_2+\widehat{rr}_3&=0,\\ \widehat{r\vartheta}_1+\widehat{r\vartheta}_2+\widehat{r\vartheta}_3&=0,\\ \widehat{r\varphi}_1+\widehat{r\varphi}_2+\widehat{r\varphi}_3&=0, \end{aligned}\right\} \quad \text{at } r=a, \tag{2.3.9}$$

and regularity at $r=0$ requires us to choose j_l for z_l. If we examine expressions (2.3.6–8), noting that the boundary conditions must be satisfied at all points on the surface, that is, for all values of ϑ and φ, we see that (2.3.9) split into two independent groups

$$\left.\begin{aligned} &(1) \quad \widehat{rr}_2=0, && \widehat{r\vartheta}_2=0, && \widehat{r\varphi}_2=0,\\ &(2) \quad \widehat{rr}_1+\widehat{rr}_3=0, && \widehat{r\vartheta}_1+\widehat{r\vartheta}_3=0, && \widehat{r\varphi}_1+\widehat{r\varphi}_3=0, \end{aligned}\right\} \tag{2.3.10}$$

and that in both groups the second and the third equations give identical equations from the boundary conditions. Thus an elastic sphere has two kinds of free oscillation. The first was called by Lamb (1882) a 'Vibration of the First Class'. In it, since $u=0$, the displacements at a point are always orthogonal to the radius to that point, and points of a spherical surface with centre at the origin remain on that surface. Moreover $\Delta=0$, so that there is no dilatation. Such a motion was at one time called 'torsional', since in simple cases the motion could be regarded as compounded of twists about an axis. However, no external torque is envisaged, and the First Class is now called '*toroidal*'.

The other kind of free oscillation was called by Lamb a 'Vibration of the Second Class'. In this, since $u\neq 0$, spherical surfaces are deformed. In a simple example, where there is axial symmetry, the deformation is to a spheroidal shape and there is no twisting. The name *spheroidal* has consequently been used for the whole of the Second Class, though some authors call it 'poloidal'. As will be seen later, the distortion may be to much more elaborate shapes than spheroids.

The boundary condition for (1) is

$$\frac{\mathrm{d}}{\mathrm{d}\eta}\{j_l(\eta)/\eta\}=0, \tag{2.3.11}$$

where

$$\eta=ka=\frac{\omega a}{c_S}=\frac{2\pi a}{c_S T}=\frac{1}{T}\frac{2\pi a}{c_S}, \tag{2.3.12}$$

T being the period corresponding to frequency ω. We observe that η is a non-dimensional frequency expressed in units of $c_S/2\pi a$, that is, η^{-1} is a non-dimensional period expressed in units of $2\pi a/c_S$, the time which a shear wave would take to travel around a full surface meridian. (This should not be confused with the time taken by a Love wave.)

Consistency of $A_{ml} : C_{ml}$ in the second group of equations (2.3.10), gives

$$2\frac{\xi}{\eta}\left[\frac{1}{\eta}+\frac{(l-1)(l+2)}{\eta^2}\left\{\frac{j_{l+1}(\eta)}{j_l(\eta)}-\frac{l+1}{\eta}\right\}\right]j_{l+1}(\xi)$$
$$+\left[-\frac{1}{2}+\frac{(l-1)(2l+1)}{\eta^2}+\frac{1}{\eta}\left\{1-\frac{2l(l-1)(l+2)}{\eta^2}\right\}\frac{j_{l+1}(\eta)}{j_l(\eta)}\right]j_l(\xi)=0, \tag{2.3.13}$$

where

$$\xi = ha = \frac{\omega a}{c_P} = \frac{1}{T}\frac{2\pi a}{c_S}\frac{c_S}{c_P} = \eta\frac{c_S}{c_P}. \tag{2.3.14}$$

The fact that only η (and so c_S) occurs in (2.3.11), but both η and ξ (and so c_S and c_P) occur in (2.3.13), shows that S and P waves are coupled in spheroidal oscillations, whereas only S waves are found in toroidal oscillations. For a uniform sphere, this coupling is due to the surface boundary conditions.

Both (2.3.11) and (2.3.13) are equations for the unknown η with parameter l, and are therefore frequency equations. When we solve numerically one of these equations with a fixed value of l, an infinite number of roots η will be obtained. Let us denote these by ${}_n\eta_l$; let $n = 0$ correspond to the minimum root of η and let the roots be numbered in ascending order of magnitude, so that n increases as the value of η increases; n takes the values

$$n = 0, 1, 2, \ldots. \tag{2.3.15}$$

The *fundamental* mode for given l is that with $n = 0$, the others are *overtones*.

It should be noticed that the frequency equations do not include m, that is, the non-dimensional frequency ${}_n\eta_l$ is independent of m. From the simple explanation in Chapter 1, an eigenfrequency of a

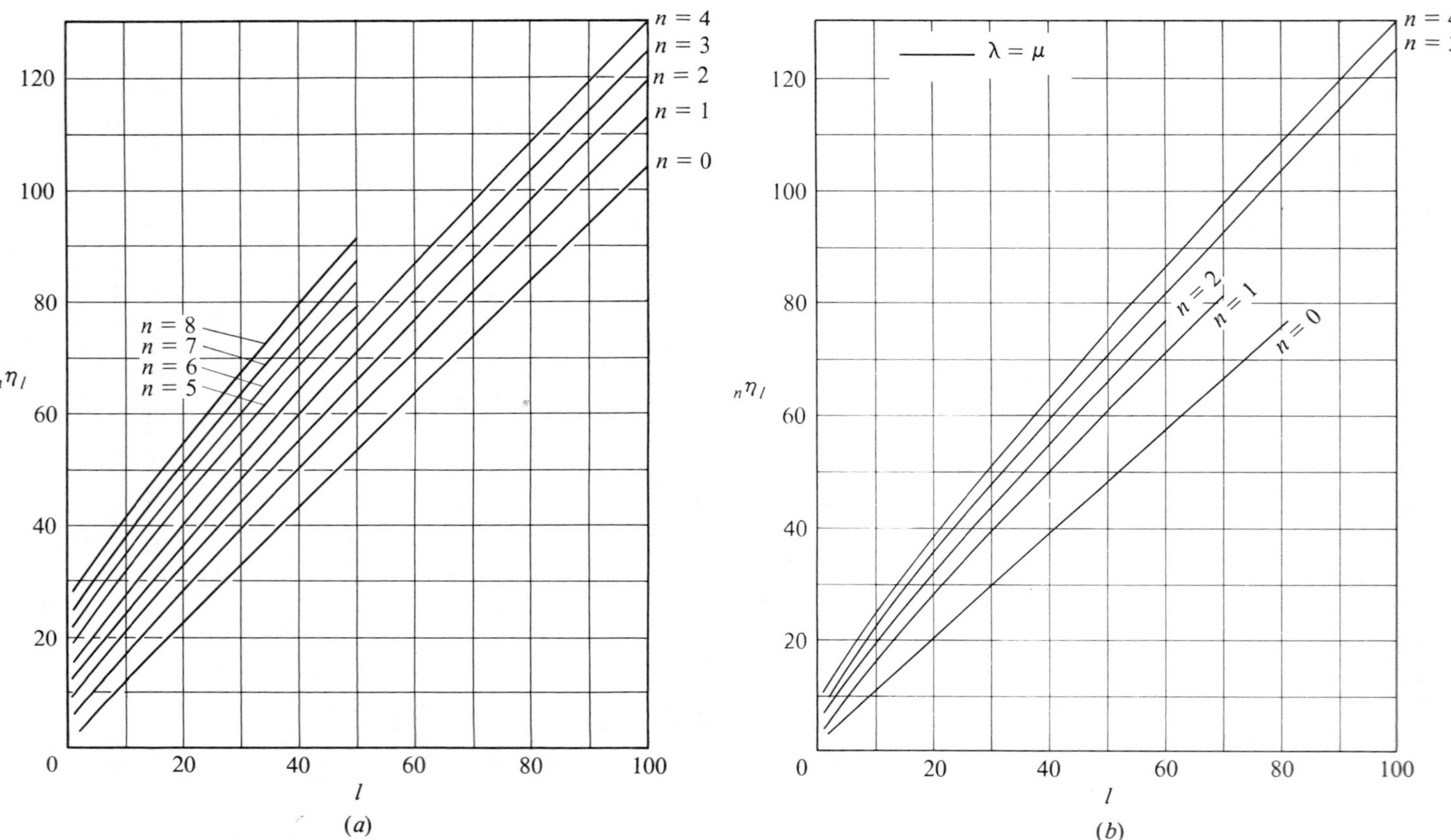

n = 4
n = 3
n = 2
n = 1
n = 0
n = 8
n = 7
n = 6
n = 5
${}_n\eta_l$
0
20
40
60
80
100
120
l
(a)
λ = μ
n = 4
n = 3
n = 2
n = 1
n = 0
${}_n\eta_l$
0
20
40
60
80
100
120
l
(b)

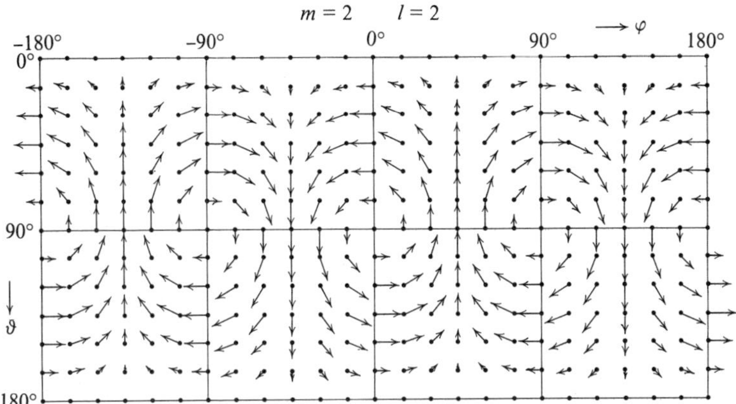

Fig. 2.5. Distribution of displacements in toroidal oscillation ${}_0T_2^2(l = 2, m = 2, n = 0)$ on the surface of a uniform sphere. Displacements lie on the surface with direction and magnitude shown by the arrows.

finite three-dimensional elastic body is expected to have three suffixes m, l and n. But this is not the case here, which implies degeneration like that explained in Chapter 1. We shall return to this point later (Chapter 7) when we introduce the effect of rotation of the Earth.

Since the expressions in (2.3.11) and (2.3.13) are not complicated, the numerical values of the roots ${}_n\eta_l$ may be calculated by the use of a computer without modifying the expressions. However, for larger values of l and η, the structure of the pattern of eigenfrequencies is more clearly shown by the use of approximate formulae for spherical Bessel functions. The choice of approximation differs according to the relation between the degree (l) and the argument (η). The asymptotic distribution of eigenfrequencies of overtones in toroidal oscillations is discussed in Chapter 8. For spheroidal oscillations Brune (1966) and Odaka (1978) have given approximate formulae for frequency equations when l and η are large, in the form

$$\cos\{\xi - \tfrac{1}{2}\pi(l+1)\}\cos\{\eta - \tfrac{1}{2}\pi(l+1)\} = 0, \qquad (2.3.16)$$

when $\eta \gg l$. This formula holds for short period higher overtone modes. We obtain two independent sets of solutions, namely, ${}_n\xi_l$ and ${}_n\eta_l$, which correspond to P and S waves respectively. This is an example of the decoupling of P and S waves which occurs in spheroi-

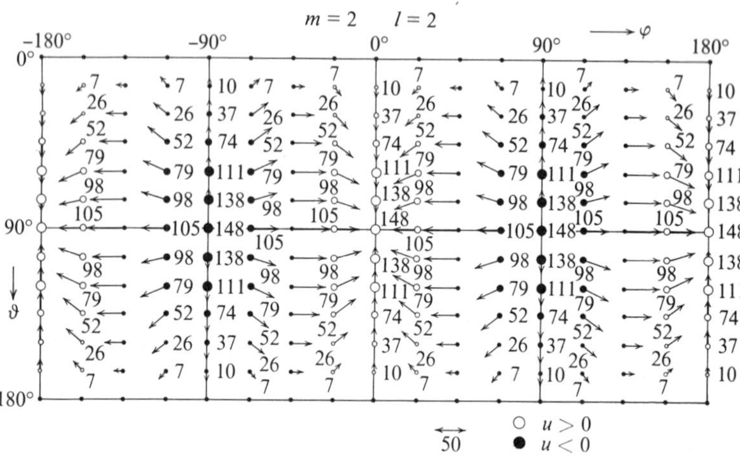

Fig. 2.6. Distribution of displacements in spheroidal oscillation ${}_0S_2^2 (l = 2, m = 2, n = 0)$ on the surface of a uniform sphere. Numbers beside points indicate radial components (open circle: up, solid circle: down). Arrows indicate direction and magnitude of tangential components.

dal oscillations at short period. Values of the non-dimensional frequency ${}_n\eta_l$ are shown in Fig. 2.4 (Satô & Usami, 1962). It is to be noticed that one mode ($l = 1, n = 0$) is missing for both toroidal and spheroidal oscillations (see § 2.4).

Figs. 2.5 and 2.6 compare examples of the patterns of displacements in free oscillations of toroidal and spheroidal types. Fig. 2.5 shows the pattern of displacement vectors on the surface for the mode ($m = 2, l = 2$) in toroidal oscillations (Satô & Usami, 1962b). There is no component of displacement orthogonal to the surface. The vectors drawn on the (φ, ϑ) plane show clearly a rotational pattern.

It is harder to represent a spheroidal distortion. In Fig. 2.6 the arrow from a point indicates direction and magnitude of the displacement orthogonal to the radius at that point. The number alongside indicates the radial component of displacement. Open circles refer to upward displacements and black circles to downward. The equator ($\vartheta = 90°$) is distorted into an ellipse with major axis at $\varphi = 0°$, $\varphi = 180°$ and minor axis at $\varphi = -90°$, $\varphi = +90°$. On any meridian the radial displacement has the same sign at all points, being greatest at the equator.

2.4 Examination of special modes for which $l = 0$ or 1

In this section we explain the character of the modes with $l = 0$ or 1, inferred from the formal expressions of the solutions $\mathbf{u}_1, \mathbf{u}_2, \mathbf{u}_3$ ((2.2.28–30)).

Toroidal oscillation Only the displacement $\mathbf{u}_2$ arises in toroidal oscillations. From the form of the Associated Legendre function $P_l^m(\cos\vartheta)$ we note that m cannot be greater than l. If $l = 0$ we have $m = 0$ and $P_0^0 = 1$. Hence $u_2 = v_2 = w_2 = 0$; the solution is trivial and no eigenfrequency is obtained.

If $l = 1$, the mode corresponding to $n = 0$ (the fundamental mode) is again missing. We show this as follows: since $l = 1$, m can be either 0 or 1. If $m = 0$, $P_1^0 = \cos\vartheta$ and $v_2 = 0$. Therefore only the φ-component of displacement w_2 exists, being proportional to $\sin\vartheta$. Each thin spherical shell of thickness dr rotates around the z-axis like a rigid shell. If $m = 1$, $P_1^1 = \sin\vartheta$ and we have

$$v_2 \propto {}^{\sin}_{\cos}\,\varphi, \qquad w_2 \propto \cos\vartheta \,{}^{\cos}_{(-\sin)}\,\varphi. \tag{2.4.1}$$

In this case a thin spherical shell rotates around the x- or y-axis, as we take the upper or lower φ-functions, as a rigid shell. From the analogy of modes for $l \geq 2$ (Fig. 2.7), this fundamental mode ($n = 0$), if it exists, can have no node in the radial distribution of displacement. This means that, for this mode, every part of the sphere shows rotation around some axis in one direction. Such a motion could not be sustained without the application of an external torque and so would not be a free oscillation. Consequently, the frequency equation has no root corresponding to this mode.

Spheroidal oscillation For spheroidal oscillations the displacement is expressed by the sum of corresponding components of $\mathbf{u}_1$ and $\mathbf{u}_3$. If $l = 0$, we have $m = 0$, $P_0^0 = 1$. Therefore v and w vanish and

$$u = u_1 = Aj_0(hr), \qquad 0 < r < a. \tag{2.4.2}$$

u takes the same value at each point of a concentric spherical surface of radius r. This is an oscillation consisting of simple dilatation and contraction about the centre of the sphere.

If $l = 1$, m must be either 0 or 1, and a mode corresponding to $n = 0$

is missing for the following reasons. When $m=0$, we have

$$\left.\begin{aligned} u(r) &\propto A\cos\vartheta,\\ v(r) &\propto -B\sin\vartheta,\\ w(r) &= 0,\end{aligned}\right\} \tag{2.4.3}$$

where A and B are functions of r. From Fig. 2.8, showing the radial distribution of eigenfunctions, we see that if we extrapolate to a mode with $l=1$ and $n=0$, A and B must be quantities of the same sign. Without loss of generality, we suppose that A and B are positive; then every part of the sphere has displacement component in the direction of positive z. This could not be sustained without the application of an external force in the positive z-direction, so the frequency equation has no root corresponding to $l=1, m=0$. In the same way, when $m=1$, we have

$$\left.\begin{aligned} u(r) &\propto -A\sin\vartheta\,{}^{\cos}_{\sin}\,\varphi,\\ v(r) &\propto -B\cos\vartheta\,{}^{\cos}_{\sin}\,\varphi,\\ w(r) &\propto B\,{}^{\ \sin}_{(-\cos)}\,\varphi.\end{aligned}\right\} \tag{2.4.4}$$

Here too we may suppose that A and B have the same sign. If the sign is positive, all parts of the sphere have displacement components in the same (negative x or y) direction according as we take the upper or lower φ-function. This cannot be a free oscillation.

When $l=1$ and the radial distance r tends to zero, the spherical Bessel functions in the expressions for displacement (2.2.28) and (2.2.30) are expanded as

$$\left.\begin{aligned} j_1(hr) &= \frac{1}{3}hr - \frac{1}{30}(hr)^3 + O\{(hr)^5\},\\ \frac{\mathrm{d}}{\mathrm{d}r}j_1(hr) &= \frac{1}{3}h - \frac{1}{10}h^3r^2 + O(h^5r^4).\end{aligned}\right\} \tag{2.4.5}$$

For the displacement $\mathbf{u}_3$, k replaces h. The form, when r is small, of displacement components of $\mathbf{u}_1$ and $\mathbf{u}_3$ is shown in Table 2.1. Consider the displacement on a spherical surface of small radius, say $r=\varepsilon$, and then let ε tend to zero. Noting that the displacement is the sum of contributions from $\mathbf{u}_1$ and $\mathbf{u}_3$, we find that when $m=0$, the centre of the sphere moves in the z-direction by an amount

Table 2.1. *Form of displacement in* $\mathbf{u}_1$ *and* $\mathbf{u}_3$ *for spheroidal oscillations when hr and kr are small.*

Solution	$m = 0$	$m = 1$
u_1	$-A\left(\frac{h}{3}-\frac{h^3r^2}{10}\right)\cos\vartheta$	$-A\left(\frac{h}{3}-\frac{h^3r^2}{10}\right)\sin\vartheta\,{}^{\cos}_{\sin}\,\varphi$
v_1	$-A\left(\frac{h}{3}-\frac{h^3r^2}{30}\right)(-\sin\vartheta)$	$-A\left(\frac{h}{3}-\frac{h^3r^2}{30}\right)\cos\vartheta\,{}^{\cos}_{\sin}\,\varphi$
w_1	0	$A\left(\frac{h}{3}-\frac{h^3r^2}{30}\right){}^{\sin}_{(-\cos)}\,\varphi$
u_3	$2C\left(\frac{1}{3}-\frac{k^2r^2}{30}\right)\cos\vartheta$	$2C\left(\frac{1}{3}-\frac{k^2r^2}{30}\right)\sin\vartheta\,{}^{\cos}_{\sin}\,\varphi$
v_3	$C\left(\frac{2}{3}-\frac{2k^2r^2}{15}\right)(-\sin\vartheta)$	$C\left(\frac{2}{3}-\frac{2k^2r^2}{15}\right)\cos\vartheta\,{}^{\cos}_{\sin}\,\varphi$
w_3	0	$-C\left(\frac{2}{3}-\frac{2k^2r^2}{15}\right){}^{\sin}_{(-\cos)}\,\varphi$

$\frac{1}{3}(-Ah+2C)$. When $m = 1$, it can be proved that the centre moves by the amount $\frac{1}{3}(-Ah+2C)$ in the x- or y-direction, according as we take the upper or lower φ-function. The motion seems strange at first sight. But let us consider the movement of the centre of mass. The displacements in the x-, y- and z-directions are

$$\left.\begin{aligned} u_x &= u\sin\vartheta\cos\varphi + v\cos\vartheta\cos\varphi - w\sin\varphi,\\ u_y &= u\sin\vartheta\sin\varphi + v\cos\vartheta\sin\varphi + w\cos\varphi,\\ u_z &= u\cos\vartheta - v\sin\vartheta. \end{aligned}\right\} \tag{2.4.6}$$

The movement of the centre of mass in the x-, y- and z-directions is proportional to

$$\int_0^a r^2\,\mathrm{d}r\int_0^\pi \sin\vartheta\,\mathrm{d}\vartheta\int_0^{2\pi}\mathrm{d}\varphi\begin{Bmatrix} u_x r\sin\vartheta\cos\varphi\\ u_y r\sin\vartheta\sin\varphi\\ u_z r\cos\vartheta\end{Bmatrix}, \tag{2.4.7}$$

which is easily proved to be zero for $m = 0$ and 1. Thus the centre of mass does not move; this explains why the movement of the centre does not contradict the physical requirement that the centre of mass stays unmoved. However, the physical reason why the geometrical centre moves when $l = 1$ is still to be clarified.

2.5 Eigenfunctions and the physical meaning of the suffixes *m*, *l* and *n*

An eigenfunction is a function which satisfies the boundary conditions and corresponds to a particular eigenfrequency. For bodies with spherical symmetry the frequency equation is independent of ϑ and φ, and so the radial eigenfunction shows the radial distribution of displacement. It will be noticed that this eigenfunction is not affected by l or m.

For toroidal oscillations of a uniform sphere, the displacement is given (from (2.2.29)) by

$$\left.\begin{aligned} u &= 0, \\ v &= mB_{ml} W_l(r) \frac{1}{\sin\vartheta} P_{l}^{m} {}_{(-\cos)}^{\ \ \sin} m\varphi \, \mathrm{e}^{\mathrm{i}\omega t}, \\ w &= B_{ml} W_l(r) \partial_\vartheta P_l^m {}_{\sin}^{\cos} m\varphi \, \mathrm{e}^{\mathrm{i}\omega t}; \end{aligned}\right\} \qquad (2.5.1)$$

for spheroidal oscillations, from (2.2.28) with (2.3.30), by

$$\left.\begin{aligned} u &= A_{ml} U_l(r) P_l^m {}_{\sin}^{\cos} m\varphi \, \mathrm{e}^{\mathrm{i}\omega t}, \\ v &= A_{ml} V_l(r) \partial_\vartheta P_l^m {}_{\sin}^{\cos} m\varphi \, \mathrm{e}^{\mathrm{i}\omega t}, \\ w &= -mA_{ml} V_l(r) \frac{1}{\sin\vartheta} P_{l}^{m} {}_{(-\cos)}^{\ \ \sin} m\varphi \, \mathrm{e}^{\mathrm{i}\omega t}, \end{aligned}\right\} \qquad (2.5.2)$$

where U, V and W are derived from spherical Bessel functions. In this book we designate U_l, V_l and W_l in (2.5.1) and (2.5.2) as eigenfunctions; U_l is the radial component and V_l, W_l the tangential components. The spheroidal oscillation has both radial and tangential components of each eigenfunction.

Figs. 2.7 and 2.8 show the functional form of $W(r)$, $U(r)$ and $V(r)$ for different values of l and n. For toroidal oscillations the surface value of $W(r)$ is standardised to 1; for spheroidal oscillations the surface value of $U(r)$ is taken to be 1. In Fig. 2.8, which shows spheroidal oscillations, the solid curves represent $U(r)$ and the dashed curves $lV(r)$. (We choose $lV(r)$ instead of $V(r)$ because, when l

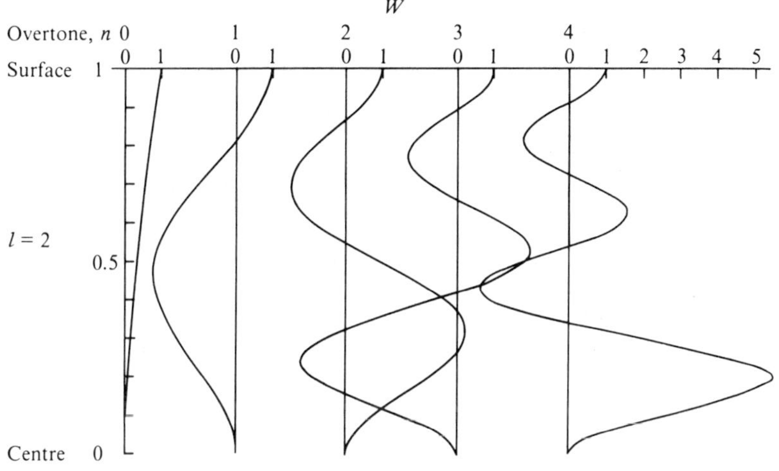
W
Overtone, n 0
1
2
3
4
0 1 0 1 0 1 0 1 0 1 2 3 4 5
Surface 1
l = 2
0.5
Centre 0

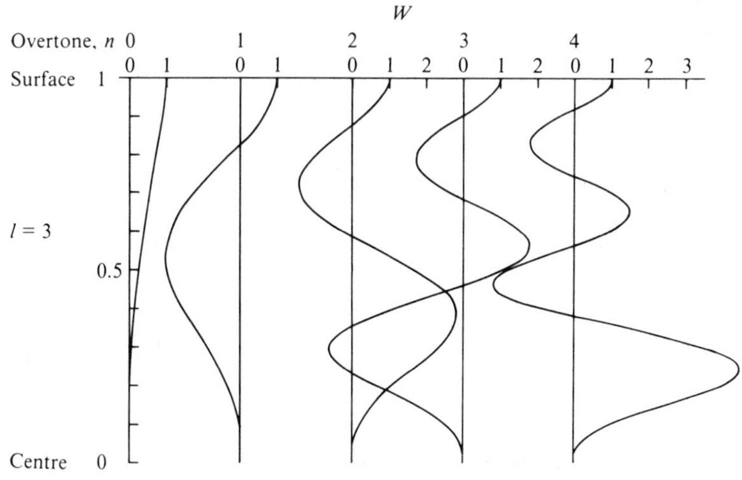
W
Overtone, n 0
1
2
3
4
0 1 0 1 0 1 2 0 1 2 0 1 2 3
Surface 1
l = 3
0.5
Centre 0

(*Cont.*)

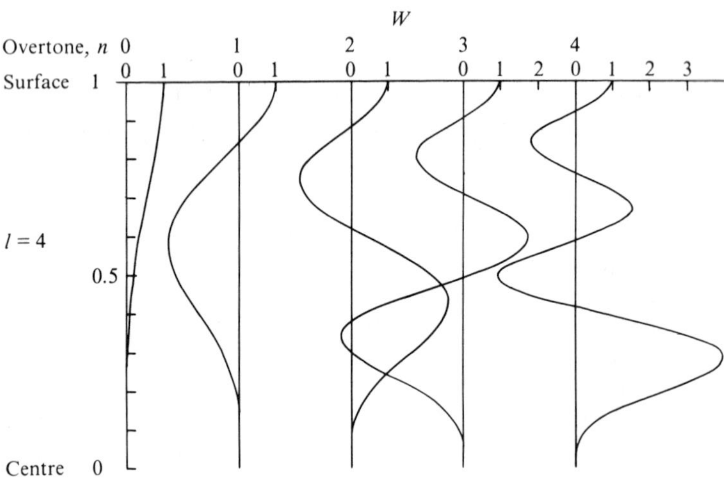

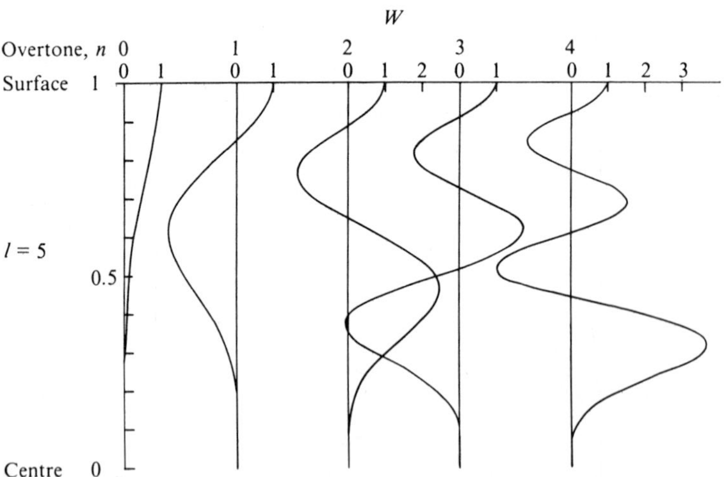

Fig. 2.7. Eigenfunctions $W(r)$ of fundamental $(n = 0)$ and four overtones $(n = 1, 2, 3, 4)$ in toroidal oscillations T_2^0, T_3^0, T_4^0, T_5^0 of a uniform sphere. The surface value is standardised at unity.

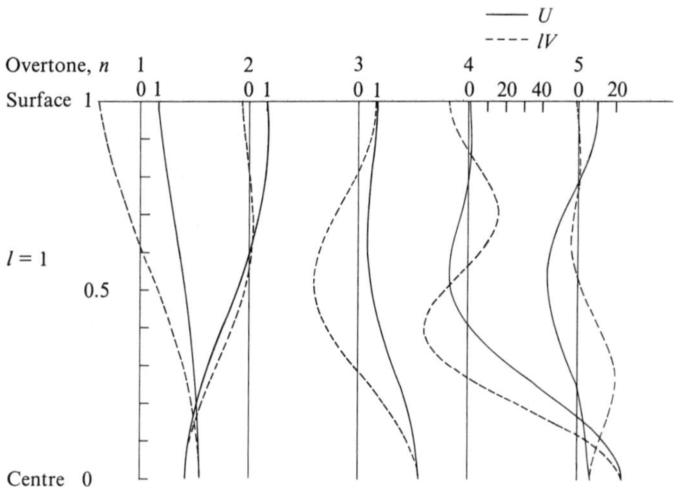
U
lV
Overtone, n 1 2 3 4 5
0 1 0 1 0 1 0 20 40 0 20
Surface 1
l = 1
0.5
Centre 0

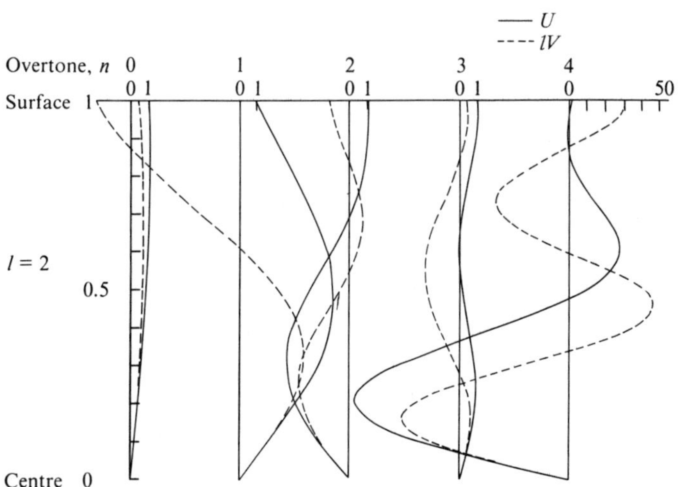
U
lV
Overtone, n 0 1 2 3 4
0 1 0 1 0 1 0 1 0 50
Surface 1
l = 2
0.5
Centre 0

(*Cont.*)

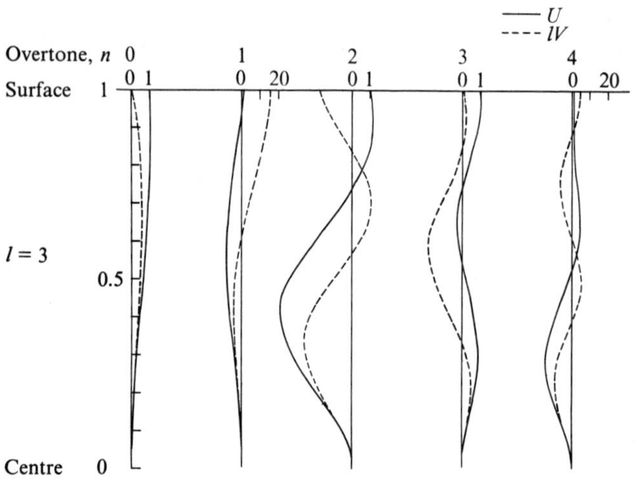

— U
---- lV
Overtone, n 0 1 2 3 4
0 1
0 20
0 1
0 1
0 20
Surface 1
l = 3
0.5
Centre 0

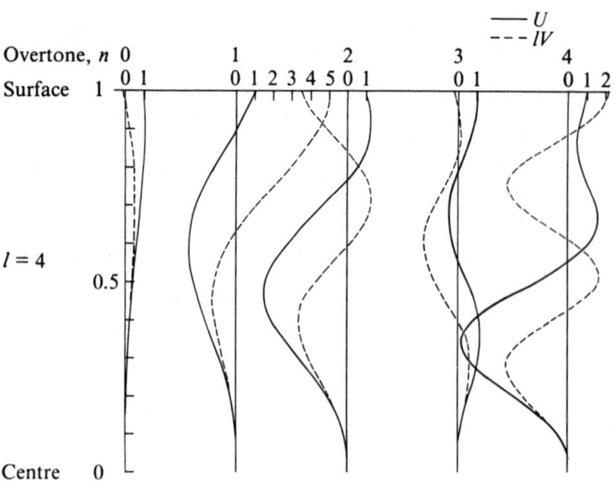

— U
--- lV
Overtone, n 0 1 2 3 4
0 1
0 1 2 3 4 5
0 1
0 1
0 1 2
Surface 1
l = 4
0.5
Centre 0

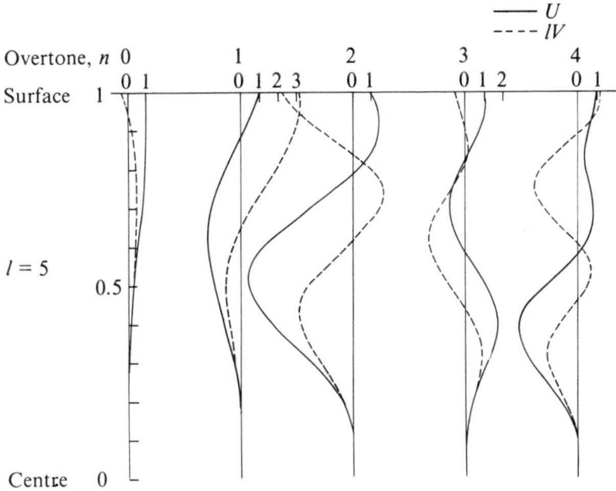

Fig. 2.8. Eigenfunctions $U(r)$ and $V(r)$ of fundamental and four overtones ($n = 0, 1, 2, 3, 4$) in spheroidal oscillations $S_2^0, S_3^0, S_4^0, S_5^0$ and five overtones ($n = 1, 2, 3, 4, 5$) in S_1^0 for a uniform sphere in which $\lambda = \mu$. Radial displacement $U(r)$ (solid curve) is standardised to unity at surface. Tangential component $lV(r)$ is shown by dashed curve.

tends to infinity, $\{lV(r)/U(r)\}_{r=a}$ tends to the well-known ratio of the horizontal and vertical displacements of plane Rayleigh waves on a plane surface. This relation will be proved later.)

The parameter m appears in the functions of ϑ and φ. It is clear from the functional forms $\sin m\varphi$ and $\cos m\varphi$ that m is the wave-number in the φ-direction. As stated earlier (in (2.3.13)), the eigen-frequency is independent of m. This will be shown to imply that the free oscillation of a uniform elastic sphere is an example of de-generation from that of a more complicated elastic sphere including the effects of rotation, ellipticity and horizontal heterogeneity.

The suffix l is the degree of the spherical surface harmonic. When $l \gg m$, we have asymptotically (Jeffreys & Jeffreys, 1956, § 24.15)

$$P_l^m(\cos\vartheta) \simeq l^m\left(\frac{2}{l\pi\sin\vartheta}\right)^{1/2}\cos\{(l+\tfrac{1}{2})\vartheta - \tfrac{1}{2}m\pi - \tfrac{1}{4}\pi\}. \quad (2.5.3)$$

This relation holds for values of ϑ which are not too close to 0 or π. If we insert the factor $\cos\omega t$ we obtain an expression for standing waves. If we multiply by exp $i\omega t$ and express in terms of exponentials we obtain from (2.5.3) the sum of travelling waves in the positive and negative ϑ-directions.

The first term of the argument of the cosine in (2.5.3.) is $(l+\frac{1}{2})\vartheta$, that is,

$$\{(l+\tfrac{1}{2})/a\}a\vartheta, \tag{2.5.4.}$$

where $a\vartheta$ is the distance along the meridian from the pole. Hence the wave-length L along the surface meridian is

$$L = 2\pi/\{(l+\tfrac{1}{2})/a\} = 2\pi a/(l+\tfrac{1}{2}). \tag{2.5.5.}$$

Since the period T is $2\pi/\omega$ and

$$\omega = \eta c_S/a \tag{2.5.6.}$$

for a progressive wave, the phase-velocity seen on the surface is

$$c = L/T = \{2\pi a/(l+\tfrac{1}{2})\}/(2\pi/\omega) = \eta c_S/(l+\tfrac{1}{2}). \tag{2.5.7.}$$

Thus the surface phase-velocity in units of S-wave velocity is

$$c/c_S = \eta/(l+\tfrac{1}{2}). \tag{2.5.8.}$$

This important formula is due to Jeans (1923). From the definition of group-velocity, the corresponding group-velocity is

$$c_g = d\omega/dk = \left(\frac{c_S}{a}d\eta\right)\bigg/\left(\frac{1}{a}dl\right) = c_S\, d\eta/dl, \tag{2.5.9.}$$

where k is the relevant wave-number $2\pi/L(=(l+\frac{1}{2})/a)$. The group-velocity corresponding to the surface phase-velocity, measured in units of c_S, is

$$c_g/c_S = d\eta/dl. \tag{2.5.10}$$

Thus, by employing the non-dimensional frequency η, we obtain the phase- and group-velocities for each mode by simple calculations, *for large values of l*. On a figure showing η as function of l, with parameter n, the phase-velocity c is approximately given, for large l, by the slope of a line connecting a point on the curve of η and the origin of the η–l diagram, and the group-velocity by the slope of the tangent to the curve $\eta(l)$ at a corresponding point (see Fig. 2.9).

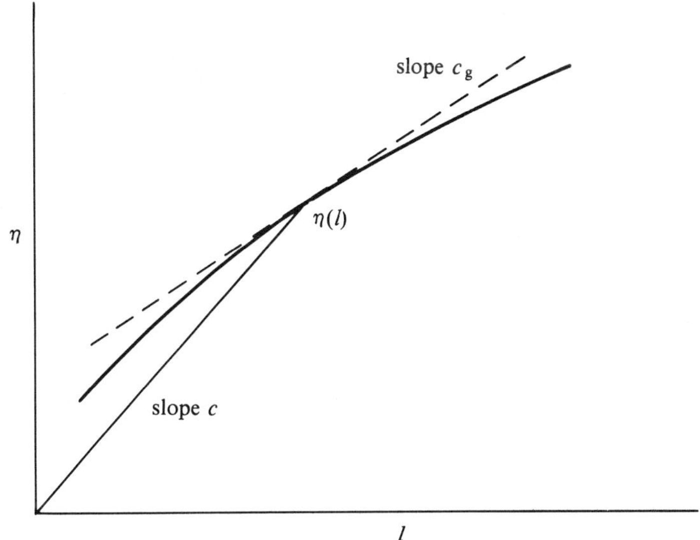

Fig. 2.9. Non-dimensional frequency η as a function of degree l.

For spheroidal oscillations the frequency equation (2.3.13) will be shown to tend to the characteristic equation of plane Rayleigh waves when l tends to infinity. From Fig. 2.8 we expect that the effective motion of the fundamental mode ($n = 0$) will be compressed into the shallower part of the sphere as l increases. We will demonstrate that this mode has a close relation to surface waves (Satô & Usami, 1962b). Large values of l correspond to short waves, and for these the curvature of the earth may be neglected. From Fig. 2.4(b) we may put

$$l > \eta(= ka) > \xi(= ha) \tag{2.5.11}$$

for the fundamental mode. Now let us substitute ν, α, β for l, ξ, η, defining the new variables by

$$\nu = l + \tfrac{1}{2}, \tag{2.5.12}$$

$$\nu \operatorname{sech} \alpha = \xi, \qquad \nu \operatorname{sech} \beta = \eta. \tag{2.5.13}$$

If $\nu \gg 1$, we have (from Jeffreys & Jeffreys, 1956, § 21.07)

$$J_\nu(\nu \operatorname{sech} \alpha) \simeq \exp\{\nu(\tanh \alpha - \alpha)\}(2\pi\nu \tanh \alpha)^{-1/2}. \tag{2.5.14}$$

After considerable manipulation we have

$$J_{\nu+1}(\nu \operatorname{sech} \alpha) \simeq \frac{\exp\{(\nu+1)(\tanh\alpha - \alpha) + \coth\alpha(\operatorname{sech}^2\alpha - 1)\}}{(2\pi)^{1/2}\{(\nu+1)\tanh\alpha + \operatorname{sech}^2\alpha \coth\alpha\}^{1/2}}. \tag{2.5.15}$$

Combining the two equations we have

$$J_{\nu+1}(\xi)/J_{\nu}(\xi) \simeq \{1 + (\coth^2\alpha)/\nu\}^{-1/2}\mathrm{e}^{-\alpha}. \tag{2.5.16}$$

Similarly we have

$$J_{\nu+1}(\eta)/J_{\nu}(\eta) \simeq \{1 + (\coth^2\beta)/\nu\}^{-1/2}\mathrm{e}^{-\beta}. \tag{2.5.17}$$

Putting (2.5.16) and (2.5.17) into (2.3.13), employing (2.5.13) and remembering $j_l(x) = (\pi/2x)^{1/2} J_{l+1/2}(x)$, we finally obtain, in terms of $\gamma = \eta/\nu$,

$$(2-\gamma^2)^2 - 4(1-\gamma^2)^{1/2}(1 - h^2\gamma^2/k^2)^{1/2} \simeq 0. \tag{2.5.18}$$

If we write $c_{\text{Rayleigh}}/c_{\text{S}}$ instead of γ, (2.5.18) becomes in the limit the characteristic equation of plane Rayleigh waves. Combining (2.5.8) with this equation, we obtain

c (phase-velocity of spheroidal mode along the surface) tends to
c_{Rayleigh} (velocity of plane Rayleigh wave) (2.5.19)

as ν tends to infinity. Thus the fundamental mode ($n = 0$) of the spheroidal oscillation of a uniform sphere is, in the limit, identical with the Rayleigh wave on the surface of a half-space.

In a similar way, we obtain for the ratio of the radial to latitudinal components of displacement

$$\left|\frac{U_l(a)}{lV_l(a)}\right| \to \frac{1 - \frac{1}{2}\gamma^2}{(1-\gamma^2)^{1/2}}, \tag{2.5.20}$$

as $\nu \to \infty$. Using the relation $\gamma = c_{\text{Rayleigh}}/c_{\text{S}}$, the right-hand side of (2.5.20) is exactly the ratio of the two components of displacement for a Rayleigh wave on a half-space.

For large values of l, we know from (2.5.3) that the order of magnitude of $\partial_\vartheta P_l^m$ is l times that of P_l^m, that is

$$\partial_\vartheta P_l^m \simeq l P_l^m. \tag{2.5.21}$$

Then, putting (2.5.20) and (2.5.21) into (2.5.2) we obtain the relative order of magnitude of amplitude components on the surface for the spheroidal oscillation (and not too close to the poles) as

$$\left.\begin{aligned} v &\simeq A_{ml}(U_l/l)lP_l^m {}^{\cos}_{\sin} m\varphi \simeq u, \\ w &\simeq -mA_{ml}(U_l/l)P^m_{l\,(-\cos)}{}^{\sin} m\varphi \simeq \frac{m}{l}u. \end{aligned}\right\} \tag{2.5.22}$$

If l is large compared with m, the longitudinal component w becomes negligibly small compared with the other components. Similarly, for toroidal oscillations we obtain

$$v \simeq \frac{m}{l}w \quad (u \text{ being zero}), \tag{2.5.23}$$

for large values of l.

For a fixed value of l, we obtain an infinite set of eigenfrequencies ${}_n\eta_l (n = 0, 1, 2, \dots)$. The lowest has $n = 0$, and (by definition) n increases as the eigenfrequency increases.

For toroidal oscillations we see from Fig. 2.7 that n appears to coincide with the number of nodal spheres. But for spheroidal oscillations there are two kinds of eigenfunction, corresponding to radial and tangential components. As is seen in Fig. 2.8, the relation between n and the number of nodal spheres is not simple. Recently a relation between n and the number of nodal spheres for spheroidal oscillations has been observed (Odaka & Usami, 1978). In Fig. 2.10, eigenfunctions of spheroidal oscillations for a fixed value $n = 14$ are shown for twelve values of l. The value of radial displacement on the surface, $U_l(a)$ is standardised to unity. (Note that the horizontal scale differs for different l). Solid and dashed curves represent the radial and tangential components respectively. The change of sign of $lV_l(a)$ from negative to positive as l increases is always followed by a unit increase in the number of nodal spheres of $U_l(r)$. Examples are seen between $l = 34$ and 35, 54 and 55, and 83 and 84. But there is a maximum degree l where the final change in sign of $lV_l(a)$ from negative to positive takes place. For instance, when $n = 14$, this maximum degree is 84. Let us call this number the critical degree $l_c(n)$. For l larger than $l_c(n)$ the number of nodal spheres of $U_l(r)$ is always n. This relation holds for all degrees l. For values of l less than $l_c(n)$ no definite relation was found between n and the number of

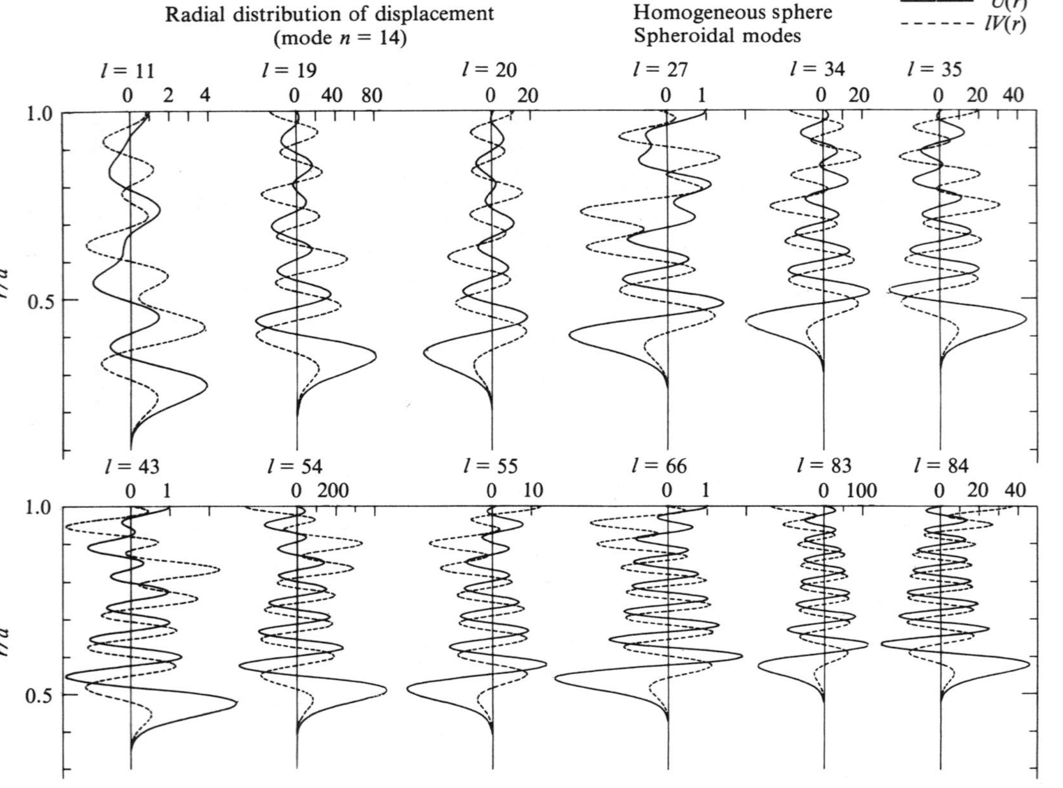

Fig. 2.10. Radial distribution of displacement $U(r)$ (solid curves) for overtone $n = 14$, for degrees $l = 11$, $19, \ldots, 84$, standardised to unity at surface. Dashed curves show $lV(r)$. Note the change in the number of

Table 2.2. *Summary of characteristics of three types of solution* $\mathbf{u}_1$, $\mathbf{u}_2$ *and* $\mathbf{u}_3$.

Displacement	Type	Derivation	Speed	Characteristics	Free oscillation	Plane wave	Predominant for large l	Limit as $l \to \infty$
$\mathbf{u}_1$	P	grad ϕ	c_P	curl $\mathbf{u}_1 = 0$ (irrotational)	Spheroidal	P	u, v	Plane Rayleigh wave (with $\mathbf{u}_3$)
$\mathbf{u}_2$	S1	curl$(r\psi, 0, 0)$	c_S	$\Delta=0, u_2=0, \widehat{rr}_2=0$ (equivoluminal)	Toroidal	SH	w	
$\mathbf{u}_3$	S2	$\frac{1}{k}$curl curl$(r\chi, 0, 0)$	c_S	$\Delta=0, (\text{curl}\,\mathbf{u}_3)_r=0$ (equivoluminal)	Spheroidal	SV	u, v	Plane Rayleigh wave (with $\mathbf{u}_1$)

nodal spheres. We observe that, for $lV_l(r)$, there is another critical degree number $l_{c1}(n)$, larger than $l_c(n)$, such that for l larger than $l_{c1}(n)$ the number of nodal spheres in $lV_l(r)$ is equal to $n+1$.

We summarise general properties of the modes as observed in the figures. It appears that:

(i) If n is fixed, the displacement is compressed into the shallower part of the sphere as l increases.

(ii) As n increases for a fixed value of l, the number of nodal spherical surfaces generally increases and the depth where the displacement becomes negligible increases.

(iii) From (i), (ii) and (2.5.19), surface waves, for large l, are formed mainly from the fundamental mode $n=0$.

(iv) From (2.5.3) the normal modes can be considered as surface waves travelling along the meridian, for large values of l and values of ϑ not close to 0 or π.

Summarising results of this section, we list the characteristics of the three forms of solution of the equations of motion $\mathbf{u}_1$, $\mathbf{u}_2$ and $\mathbf{u}_3$ in Table 2.2.

2.6 Short comments on mode-theory and ray-theory

The generation and propagation of elastic waves in a finite elastic medium has been studied by means of ray-theory and mode-theory.

From the standpoint of mode-theory, the motion of a finite elastic body can be expressed by solutions of field equations satisfying the boundary conditions, that is, by the weighed sum of contributions of all the modes. The coefficient multiplying each eigenfunction is determined by the initial conditions. In this chapter, we found from the simplest example of an elastic sphere that there are no other solutions satisfying the boundary conditions except mode solutions. Body waves propagating in the sphere should therefore be expressible in terms of modes. Mode solutions represent standing waves; yet the fact that they can express propagating waves must follow in some way from the fact that each mode may also be expressed as the sum of waves travelling to the positive and negative ϑ-directions (cf. (2.5.3.)). This leads to the following questions:

Which modes do express body waves? Which modes express particular body waves such as P, S, PP, ScS and so on? As we saw in the previous section, the fundamental modes ($n = 0$) are clearly related to surface waves; we are inevitably led to a conjecture that radial overtones ($n \geq 1$) are related to body waves. We will answer the second question in Chapter 9 where we deal with the problem of excitation of motion in layered spheres, that is, the problem of theoretical seismograms.

In ray-theory, which is essentially a short-wave approximation, the change of wave form is traced along the geometrical path from the source to the observation point. All possible rays reaching the observation point should be added to show the complete short-wave arrival at the point. Waves suffer a variety of effects such as reflection, refraction, spreading, dispersion and attenuation, and these effects must all be considered. That part of each refracted wave which propagates along boundary surfaces must also be accounted for. In ray-theory, the local change of wave form is determined mainly by the local properties of the elastic medium. On the other hand, in mode-theory a solution cannot be obtained unless the properties of whole medium are known. The detailed correspondence between the two theories is being studied by many seismologists and some interesting results have been published. For example, Odaka (1978) has succeeded in obtaining from ray-theory an asymptotic frequency equation in the short-wave approximation for the radial oscillations of a radially symmetric sphere. He has done the same for the spheroidal oscillations of a uniform elastic sphere. The basic idea is that a stationary vibration of an elastic sphere can be realised by interference of body waves travelling in the sphere.

3

TOROIDAL OSCILLATION OF A UNIFORM SHELL

The uniform sphere studied in the previous chapter is too simple to represent adequately the complexity of the Earth. Among the features to be added, the most obvious and geophysically meaningful is the liquid core – liquid since S waves through the core have never been observed. So the core may be considered to have zero rigidity ($\mu = 0$) but a finite value of λ. Only the interface conditions at the core–mantle boundary will affect toroidal oscillations; other characteristics of the core have no effect. Thus the study of toroidal oscillations of a uniform shell is relevant to that of the Earth's toroidal oscillations. On the other hand, for spheroidal oscillations the character of the core greatly influences the vibration of the whole sphere, so that the study of a uniform shell has little relation to the actual behaviour of the Earth. It is for this reason that no extensive study of the spheroidal oscillation of a uniform shell can be found.[†]

3.1 Frequency equation of toroidal oscillations of a uniform shell

Let a and b be the radii of the outer and inner surfaces of a spherical shell. In the general solution of the equations of motion for a uniform elastic medium (2.2.28–30), the spherical Bessel function $z_l(r)$ can be either $j_l(r)$ or $y_l(r)$ for the shell, since neither $j_l(r)$ nor $y_l(r)$ shows irregularities in the range $b \leq r \leq a$. Thus we have six arbitrary coefficients in the general solution.

The boundary conditions are the vanishing of stress components on the outer and inner surfaces of the shell, namely

$$\widehat{rr} = 0, \qquad \widehat{r\vartheta} = 0, \qquad \widehat{r\varphi} = 0 \qquad \text{at } r = a, b. \tag{3.1.1}$$

[†] Jobert (1957a) computed the period of the lowest spheroidal mode of a uniform shell, applying Rayleigh's Principle. She assumed that $c_S = 6.5$ km/s, $a = 6371$ km, $b = 3473$ km and $\lambda = \mu$ and obtained the period of 51.3 min. The value for a realistic Earth-model is 56.8 min, so that the error amounted to about 10%.

Thus we obtain six homogeneous equations relating the six constants. These equations split into two groups, two equations arising from the solution $\mathbf{u}_2$, and the other four from $\mathbf{u}_1$ and $\mathbf{u}_3$. The former group characterises toroidal oscillations and the latter spheroidal. The full frequency equation for toroidal oscillation is the consistency condition:

$$\begin{vmatrix} GJ(\eta) & GY(\eta) \\ GJ(\zeta) & GY(\zeta) \end{vmatrix} = 0, \tag{3.1.2}$$

where

$$\left.\begin{aligned} GJ(\eta) &= \frac{\mathrm{d}}{\mathrm{d}\eta} j_l(\eta) - \frac{1}{\eta} j_l(\eta), \\ GY(\eta) &= \frac{\mathrm{d}}{\mathrm{d}\eta} y_l(\eta) - \frac{1}{\eta} y_l(\eta), \end{aligned}\right\} \tag{3.1.3}$$

and

$$\eta = ka, \qquad \zeta = kb. \tag{3.1.4}$$

Clearly (3.1.2) could have been derived from the field equation for $\mathbf{u}_2$ alone, with one condition on each boundary. We omit consideration of spheroidal oscillations for the reason given above.

3.2 Eigenfrequencies of toroidal oscillations of a uniform shell

Matumoto & Satô (1954) were the first to solve this eigenfrequency problem, assuming that

$$\frac{b}{a} = \frac{6370-2900}{6370} \approx 0.544\,74 \tag{3.2.1}$$

and

$$c_{\mathrm{S}} = 6.5 \text{ km/s}. \tag{3.2.2}$$

They obtained the period of 42.5 min for the slowest mode ($m = 0$, $l = 2, n = 0$). They also calculated eigenfrequencies for $l \leq 4$ and $n \leq 2$. At that time, no electronic computer was available and they did the numerical computation on a hand-calculator, expanding the spherical Bessel functions in power series of the arguments. In 1956, Jobert solved the same problem for the same assumptions as

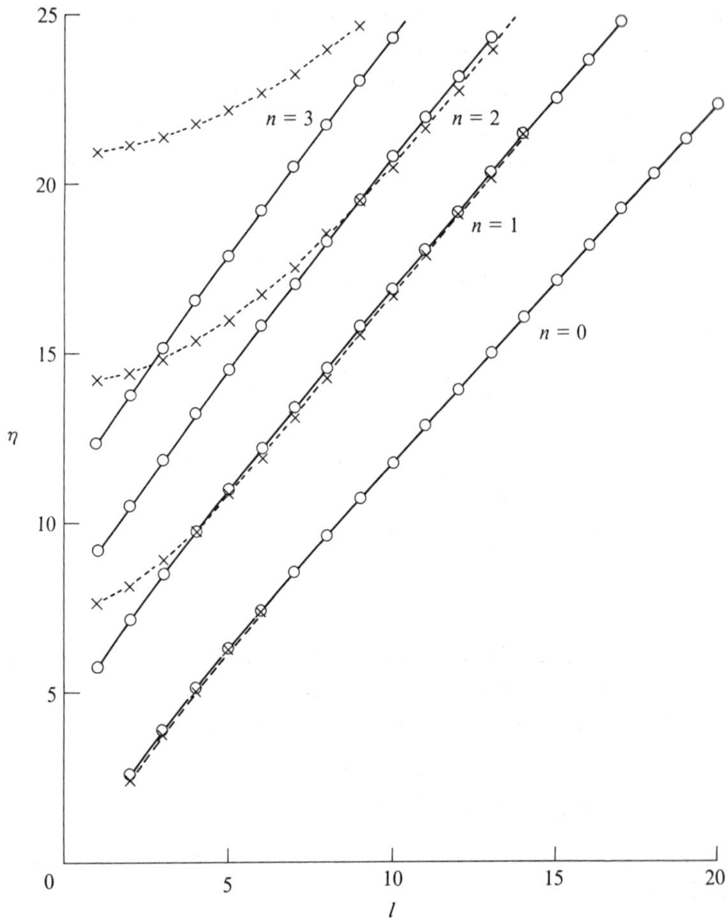

Fig. 3.1. Non-dimensional frequency $\eta(= ka = \omega a/c_S)$ of toroidal oscillations as function of l. Open circles and solid curves refer to a uniform sphere and crosses and dashed curves to a uniform shell.

to physical constants, b/a and c_S, but by using Rayleigh's Principle. She obtained a period of approximately 42.36 min. The difference between her value and that of Matumoto & Satô was 0.33%. She thus demonstrated the usefulness of Rayleigh's Principle in free oscillation studies.

Extensive computations were undertaken by Satô, Usami, Landisman & Ewing (1963), with the intention of clarifying the

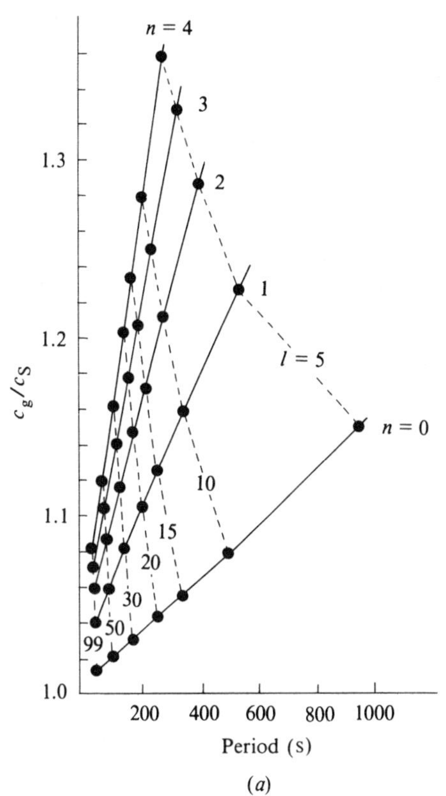

n = 4
3
2
1
l = 5
n = 0
10
15
20
30
50
99
1.3
1.2
1.1
1.0
c_g/c_S
200
400
600
800
1000
Period (s)

(a)

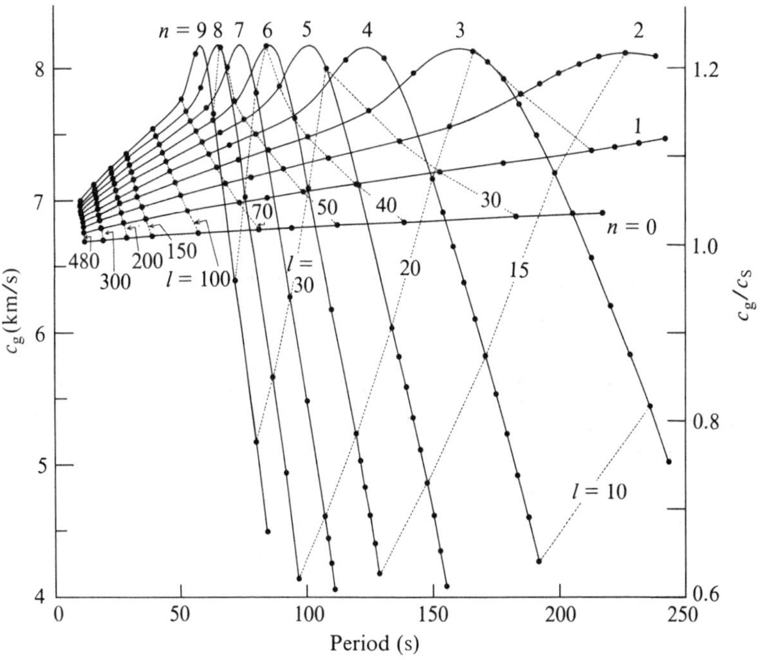

(b)

(Cont.)

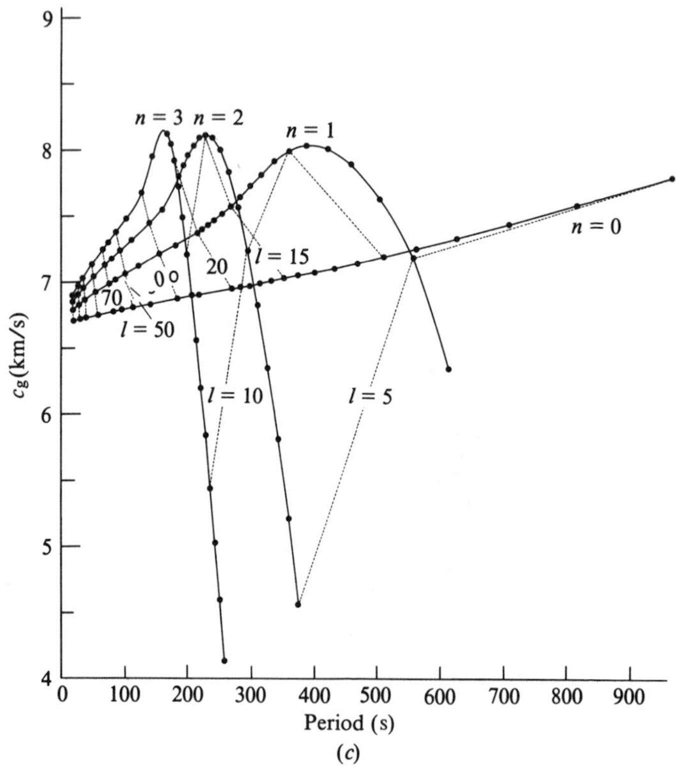

(*c*)

Fig. 3.2. Group-velocity as function of period for toroidal oscillation of (*a*) a uniform sphere and (*b*) and (*c*) a uniform shell. (*b*) and (*c*) both refer to uniform shell but their range of period (abscissa) is different. The radius of the Earth, a, is assumed to be 6370 km and $c_S = 6.5$ km/s. Hence $2\pi a/c_S = 6157$s ≈ 102.6 min.

effect of the core on the free toroidal oscillation, and of explaining from the standpoint of modal theory how the ScS wave emerges. Fig. 3.1 shows the non-dimensional frequency $\eta\,(= ka)$ as a function of Legendre degree l and overtone number n. The solid lines and open circles refer to a uniform sphere, while the broken lines and crosses refer to a uniform shell. η is independent of the absolute values of a and c_S, but depends on the ratio b/a for the spherical shell. The value of b/a is again taken as $3470/6370 \approx 0.54474$. It is important to note that although continuous curves are shown, only the points corresponding to integral values of l have physical meaning.

We see from the figure that for $n = 1$ and $n = 2$ the non-dimensional frequency of a uniform shell approaches that of a uniform sphere from below as l increases, and for large values of l the two frequencies practically coincide. Remarkable differences between the two sets of frequencies are seen for small values of l. Another difference appears in the group-velocities which have been calculated from (2.5.10). The group-velocity curves as functions of period, when the overtone n is fixed, are monotonic for the uniform sphere (see Fig. 3.2(a)). On the other hand, those for the uniform shell show a maximum, its value being nearly the same for all values of n (Fig. 3.2(b)). This maximum value is approximately $c_g \approx 1.21\ c_S$. A body wave which travels in a uniform shell, grazing the inner surface of that shell, has group-velocity $c_g = 1.18\ c_S$. The closeness of the two group-velocity values suggests a close relation between the calculated maximum group-velocity and the velocity of a wave which travels in the innermost part of the shell (Satô, Landisman & Ewing, 1960).

3.3 Eigenfunctions of toroidal oscillations of a uniform shell

Eigenfunctions calculated for some modes of the same model are shown on Fig. 3.3, where amplitude is graphed against radius for various l and n. The following features may be recognised in the figures:

(i) The amplitude at the inner surface has a non-zero value for smaller values of l and larger values of n. For these modes, the ratio of the amplitude on the inner surface to that on the outer surface increases as n increases.

(ii) For larger values of l and smaller values of n, the amplitude in the deeper part of the shell becomes practically zero. The depth below which the amplitude is negligible decreases as l increases and as n decreases. In other words, as l increases the motion is compressed into the shallower part of the shell. For these modes, the eigenfrequencies for the uniform sphere and uniform shell nearly coincide. This means that the existence of an empty core has almost no effect on the modes with higher (Legendre) degree number l.

(iii) The overtone number n equals the number of nodal spherical surfaces in the shell.

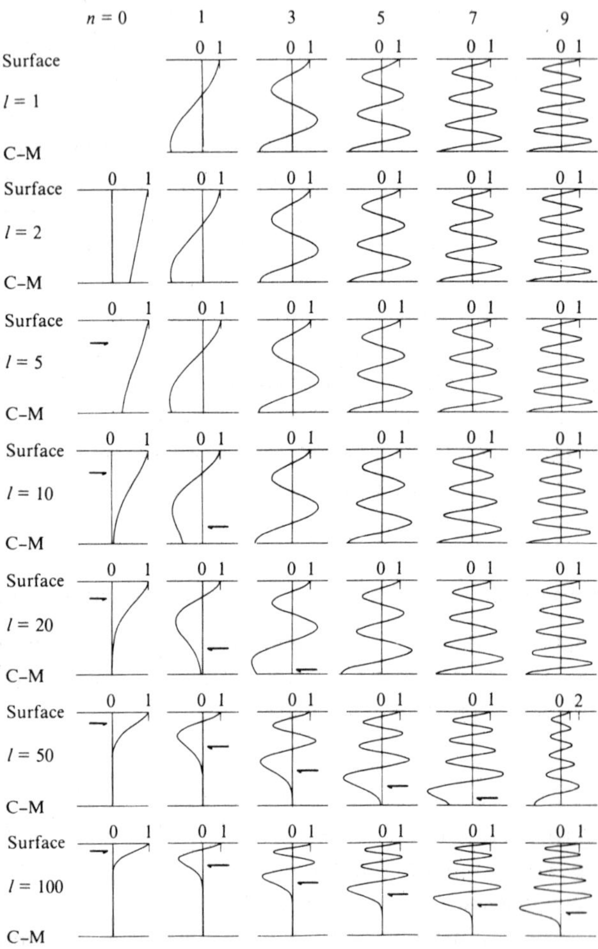

Fig. 3.3. Radial distribution of amplitude of eigenfunctions of toroidal oscillations for a uniform shell. Surface denotes the outer surface and C–M the inner surface (core–mantle boundary) of a uniform shell. The amplitude at the surface is taken as unity.

3.4 Some properties of eigenfrequencies in short period approximations

The asymptotic distribution for large overtone number n of toroidal eigenfrequencies of a uniform spherical shell was studied by Sato & Lapwood (1977a) for various relations between radial wave-length

$2\pi/k$, radii a, b and degree l:

(1) $kb \gg l$; $\quad kb \to \infty$,

(2) $kb > l + \frac{1}{2}$; $\quad kb, l \to \infty$,

(3) $ka > l + \frac{1}{2} > kb$; $\quad ka, l \to \infty$.

For (1), that is, for higher overtones, they show that the non-dimensional eigenfrequency is given approximately by

$$\eta_n \simeq \frac{n\pi a}{\gamma c_S}\left\{1 + \frac{l^2 + l + 4}{2ab}\left(\frac{a-b}{n\pi}\right)^2\right\}, \tag{3.4.1}$$

where γ is the time for a shear wave to travel radially from inner to outer surface. We see that the zero-order approximation gives $\eta_n = n\pi a/\gamma c_S$, that is, the values of η_n are equidistant at interval $\pi a/\gamma c_S$.

In (2) and (3), l is assumed to be large, so that wave-lengths along a meridian are short. Sato & Lapwood showed that the frequency equation is then the formula for constructive interference of the ScS wave (case (2)) or S wave (case (3)). This example suggests a close relation between mode- and ray-theories in the short-wave approximation (see Chapter 8).

3.5 Comments on ScS waves and normal modes

According to ray-theory the introduction of an inner free surface causes ScS to appear – an outstanding difference from the field of a uniform sphere. Can this ScS wave be expressed as the sum of contributions from modal solutions? We must look for differences of modal solutions for the uniform sphere and uniform shell in seeking an explanation of the emergence of ScS.

The eigenfrequencies do not show differences for short waves, but do display big differences for long waves; but, even for long waves, corresponding eigenfunctions show quite similar radial distributions. However, for most modes with larger n and smaller l, the amplitude of the eigenfunction is larger on the inner surface than on the outer surface of a shell. We do find a remarkable difference in the existence of maxima in the group-velocity curves as functions of period for a uniform shell; moreover, the value of those maxima is close to the group-velocity of the S wave which grazes the inner surface (calculated ray-theoretically). These graphical features suggest the following clues to the elucidation of the mode–ray relation:

(i) Modes with maximum amplitudes at depth d may have a close relation to seismic rays having the depth d as deepest point of path.

(ii) Modes and rays which have similar values of group-velocity may be closely related.

4

TOROIDAL AND SPHEROIDAL FREE OSCILLATIONS OF A TWO-LAYERED SPHERE

4.1 Frequency equations and boundary conditions

The next complexity to be introduced is the elasticity of the core. A perfect-liquid core slips at the core–mantle boundary, and so does not affect the toroidal oscillations; so a two-layered sphere with a perfect-liquid core can be replaced by a shell, which has been treated in the previous chapter. In this chapter we proceed to the general effect of a core with finite rigidity. Both core and mantle are assumed uniform.

Let a and b be the radii of sphere and core, respectively. Quantities in the mantle and core are distinguished by suffixes o and i. The boundary conditions are:

(1) vanishing of three stress components on the surface $r = a$,
(2) continuity of three stress and three displacement components on the core–mantle boundary $r = b$,
(3) regularity at the centre $r = 0$.

In the solutions (2.2.28–30) for the mantle, a linear combination of spherical Bessel functions j_l and y_l must replace z_l, while for the core, by the requirement (3) only j_l is permissible. Thus we have nine equations with nine unknown constants, say

$$A_o, A'_o, B_o, B'_o, C_o, C'_o, A_i, B_i, C_i.$$

As in the simpler systems of earlier chapters, the nine boundary equations divide into two groups. The one, giving toroidal oscillations, consists of the solution $\mathbf{u}_2$ only, while the other, relating to spheroidal oscillations, consists of solutions $\mathbf{u}_1$ and $\mathbf{u}_3$.

Toroidal oscillations The boundary conditions for toroidal oscillations are

$$\begin{aligned}
&\widehat{r\vartheta}_o = 0, && \widehat{r\varphi}_o = 0 && \text{at } r = a,\\
&\widehat{r\vartheta}_o = \widehat{r\vartheta}_i, && \widehat{r\varphi}_o = \widehat{r\varphi}_i && \\
&v_o = v_i, && w_o = w_i && \Big\} \text{ at } r = b.
\end{aligned}$$

Conditions for $\widehat{r\vartheta}$ and $\widehat{r\varphi}$ give identical equations. So do those for v and w. These six conditions therefore reduce to three homogeneous equations for B_o, B_o' and B_i, and the frequency equation is expressed in terms of $\eta_o = k_o a$, $\zeta_o = k_o b$, $\zeta_i = k_i b$ as

$$\begin{vmatrix} GJ(\eta_o) & GY(\eta_o) & 0 \\ GJ(\zeta_o) & GY(\zeta_o) & -\dfrac{k_i\mu_i}{k_o\mu_o}GJ(\zeta_i) \\ FJ(\zeta_o) & FY(\zeta_o) & -FJ(\zeta_i) \end{vmatrix} = 0, \tag{4.1.1}$$

where

$$\left.\begin{aligned}
GJ(\eta) &= \frac{d}{d\eta}j_l(\eta) - \frac{1}{\eta}j_l(\eta),\\
GY(\eta) &= \frac{d}{d\eta}y_l(\eta) - \frac{1}{\eta}y_l(\eta),\\
FJ(\zeta) &= j_l(\zeta), \qquad FY(\zeta) = y_l(\zeta).
\end{aligned}\right\} \tag{4.1.2}$$

The frequency equation (4.1.1) can be expressed as

$$FJ(\zeta_i)L_u(\eta_o,\zeta_o) - \left(\frac{\mu_i\rho_i}{\mu_o\rho_o}\right)^{1/2} GJ(\zeta_i)R_u(\eta_o,\zeta_o) = 0, \tag{4.1.3}$$

where

$$\left.\begin{aligned}
L_u(\eta_o,\zeta_o) &= GJ(\eta_o)GY(\zeta_o) - GY(\eta_o)GJ(\zeta_o)\\
R_u(\eta_o,\zeta_o) &= GJ(\eta_o)FY(\zeta_o) - GY(\eta_o)FJ(\zeta_o).
\end{aligned}\right\}, \tag{4.1.4}$$

If we put $\mu_i = 0$, or let μ_o tend to infinity, in equation (4.1.3), we obtain $L_u = 0$ or $FJ(\zeta_i) = 0$. $L_u = 0$ is identical with (3.1.2), the frequency equation for an elastic shell with a perfect-liquid core, that is, of an empty shell. It is also clear that

$$FJ(\zeta_i) = 0 \tag{4.1.5}$$

is the frequency equation for an elastic core with a rigid boundary.

When $\mu_o = 0$ or μ_i tends to infinity, we obtain $R_u(\eta_o, \zeta_o) = 0$, or $GJ(\zeta_i) = 0$. $R_u = 0$ is the frequency equation of shell vibrations where there is a rigid core, and $GJ(\zeta_i) = 0$ the frequency equation of free toroidal oscillation of an elastic core with a free boundary. Thus the frequency equation of the two-layered sphere (4.1.3) may be said to be a combination of frequency equations of the mantle and core.

Spheroidal oscillations Let $\xi_o = h_o a, \eta_o = k_o a, \chi_o = h_o b, \zeta_o = k_o b, \chi_i = h_i b, \zeta_i = k_i b$; then the frequency equation is given by

$$\begin{vmatrix} \mu_o h_o^2 D(j_l, \xi_o) & \mu_o h_o^2 D(y_l, \xi_o) & 2\mu_o l(l+1)k_o E_o(j_l, \eta_o) & 2\mu_o l(l+1)k_o E(y_l, \eta_o) & 0 & 0 \\ -2\mu_o h_o^2 E(j_l, \xi_o) & -2\mu_o h_o^2 E(y_l, \xi_o) & \mu_o k_o H(j_l, \eta_o) & \mu_o k_o H(y_l, \eta_o) & 0 & 0 \\ \mu_o h_o^2 D(j_l, \chi_o) & \mu_o h_o^2 D(y_l, \chi_o) & 2\mu_o l(l+1)k_o E(j_l, \zeta_o) & 2\mu_o l(l+1)k_o E(y_l, \zeta_o) & -\mu_i h_i^2 D(j_l, \chi_i) & -2\mu_i l(l+1)k_i E(j_l, \zeta_i) \\ -2\mu_o h_o^2 E(j_l, \chi_o) & -2\mu_o h_o^2 E(y_l, \chi_o) & \mu_o k_o H(j_l, \zeta_o) & \mu_o k_o H(y_l, \zeta_o) & 2\mu_i h_i^2 E(j_l, \chi_i) & -\mu_i k_i H(j_l, \zeta_i) \\ -h_o I(j_l, \chi_o) & -h_o I(y_l, \chi_o) & l(l+1)K(j_l, \zeta_o) & l(l+1)K(y_l, \zeta_o) & h_i I(j_l, \chi_i) & -l(l+1)K(j_l, \zeta_i) \\ -h_o K(j_l, \chi_o) & -h_o K(y_l, \chi_o) & L(j_l, \zeta_o) & L(y_l, \zeta_o) & h_i K(j_l, \chi_i) & -L(j_l, \zeta_i) \end{vmatrix} = 0 \tag{4.1.6}$$

In this equation the columns from left to right are coefficients of unknown constants $A_o, A_o', C_o, C_o', A_i, C_i$, respectively, and the rows from the top downwards represent the boundary conditions $\widehat{rr} = 0, \widehat{r\vartheta} = 0$ at $r = a$ and continuity of $\widehat{rr}$, $\widehat{r\vartheta}, u$ and v at $r = b$ respectively. The functions D, E, H, I, K, L are defined by

$$\left.\begin{aligned} D(j_l, x) &= \left(\frac{\lambda}{\mu} + 2\right) j_l(x) + \frac{4}{x}\frac{\mathrm{d}}{\mathrm{d}x} j_l(x) - \frac{2l(l+1)}{x^2} j_l(x), \\ E(j_l, x) &= \frac{1}{x}\frac{\mathrm{d}}{\mathrm{d}x} j_l(x) - \frac{1}{x^2} j_l(x), \\ H(j_l, x) &= \frac{\mathrm{d}^2}{\mathrm{d}x^2} j_l(x) + \frac{l(l+1) - 2}{x^2} j_l(x), \\ I(j_l, x) &= \frac{\mathrm{d}}{\mathrm{d}x} j_l(x), \\ K(j_l, x) &= \frac{1}{x} j_l(x), \\ L(j_l, x) &= \frac{\mathrm{d}}{\mathrm{d}x} j_l(x) + \frac{1}{x} j_l(x), \end{aligned}\right\} \tag{4.1.7}$$

and similarly for other arguments. Lamé's constants λ and μ in the expression for $D(j_l, x)$ occur in the form λ_o/μ_o and λ_i/μ_i in mantle and core, respectively. Equation (4.1.6) may be reduced to a more convenient form for numerical calculations, namely

$$\begin{vmatrix} D(j_l,\xi_o) & D(y_l,\xi_o) & 2l(l+1)E(j_l,\eta_o) & 2l(l+1)E(y_l,\eta_o) & 0 & 0 \\ -2E(j_l,\xi_o) & -2E(y_l,\xi_o) & H(j_l,\eta_o) & H(y_l,\eta_o) & 0 & 0 \\ D(j_l,\chi_o) & D(y_l,\chi_o) & 2l(l+1)E(j_l,\zeta_o) & 2l(l+1)E(y_l,\zeta_o) & -\dfrac{\mu_i}{\mu_o}D(j_l,\chi_i) & -2\dfrac{\mu_i}{\mu_o}l(l+1)E(j_l,\zeta_i) \\ -2E(j_l,\chi_o) & -2E(y_l,\chi_o) & H(j_l,\zeta_o) & H(y_l,\zeta_o) & 2\dfrac{\mu_i}{\mu_o}E(j_l,\chi_i) & -\dfrac{\mu_i}{\mu_o}H(j_l,\zeta_i) \\ -c_{Po}I(j_l,\chi_o) & -c_{Po}I(y_l,\chi_o) & c_{So}l(l+1)K(y_l,\zeta_o) & c_{So}l(l+1)K(y_l,\zeta_o) & c_{Pi}I(j_l,\chi_i) & -c_{Si}l(l+1)K(j_l,\zeta_i) \\ -c_{Po}K(j_l,\chi_o) & -c_{Po}K(y_l,\chi_o) & c_{So}L(j_l,\zeta_o) & c_{So}L(y_l,\zeta_o) & c_{Pi}K(j_l,\chi_i) & -c_{Si}L(j_l,\zeta_i) \end{vmatrix} = 0, \tag{4.1.8}$$

where c_P and c_S are velocities of P and S waves. When the core is empty, $\mu_i = 0, c_{Si} = 0, c_{Pi} = 0$, and this equation reduces to

$$\bar{\Delta}(5, 6; 5, 6) = 0. \tag{4.1.9}$$

Here $\bar{\Delta}(5, 6; 5, 6)$ represents the minor determinant obtained by eliminating the fifth and sixth columns and rows from the determinant in (4.1.8). Thus (4.1.9) is the frequency equation of spheroidal oscillation of a uniform shell.

When the core consists of perfect-liquid, the solution $\mathbf{u}_3$ in the core disappears because no S wave can travel in liquid. We then obtain the frequency equation by putting μ_i and c_{Si} zero and eliminating the last row, which is derived from the condition of continuity of tangential displacement at the core–mantle boundary, and the last column, which is contributed by $\mathbf{u}_3$.

4.2 Spheroidal oscillations of a sphere with a perfect-liquid core

The eigenfrequencies of spheroidal oscillations of a two-layered model consisting of a uniform mantle and a perfect-liquid core are obtained by solving numerically

$$\bar{\Delta}(6, 6) = 0, \tag{4.2.1}$$

which is derived from (4.1.8) by eliminating the sixth column and row and putting $\mu_i = 0$. Introducing average values of the Earth's material constants in the mantle and core, and assuming $\lambda = \mu$ in the mantle, we have adopted the following values in our numerical

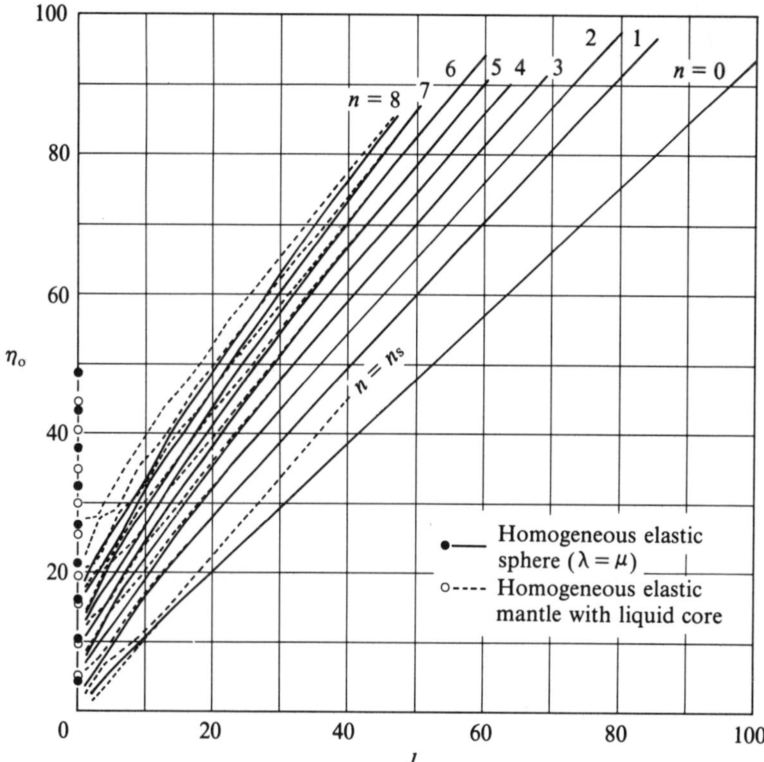

Fig. 4.1. Non-dimensional frequency $\eta_o (= k_o a)$ of the free spheroidal oscillation for a uniform sphere with a perfect-liquid core, compared with η_o for a uniform solid sphere. (n_s refers to the special branch which does not exist for a uniform sphere.)

computation:

$$\left.\begin{array}{lll} \text{radius of the sphere} & a = 6370 \text{ km}, & \\ \text{radius of the core} & b = (6370 - 2900) = 3470 \text{ km}, & \\ & (b/a = 0.54474), & \\ \text{density ratio of core to mantle} & \rho_i/\rho_o = 2.2, & \\ \text{P- and S-wave velocities} & c_{P_o} = 11.55 \text{ km/s}, & \\ & c_{S_o} = c_{P_o}/\sqrt{3} = 6.667 \text{ km/s} & \\ & c_{P_i} = 10.00 \text{ km/s}. & \end{array}\right\} \quad (4.2.2)$$

Fig. 4.1 (Satô & Usami, 1964) shows the non-dimensional fre-

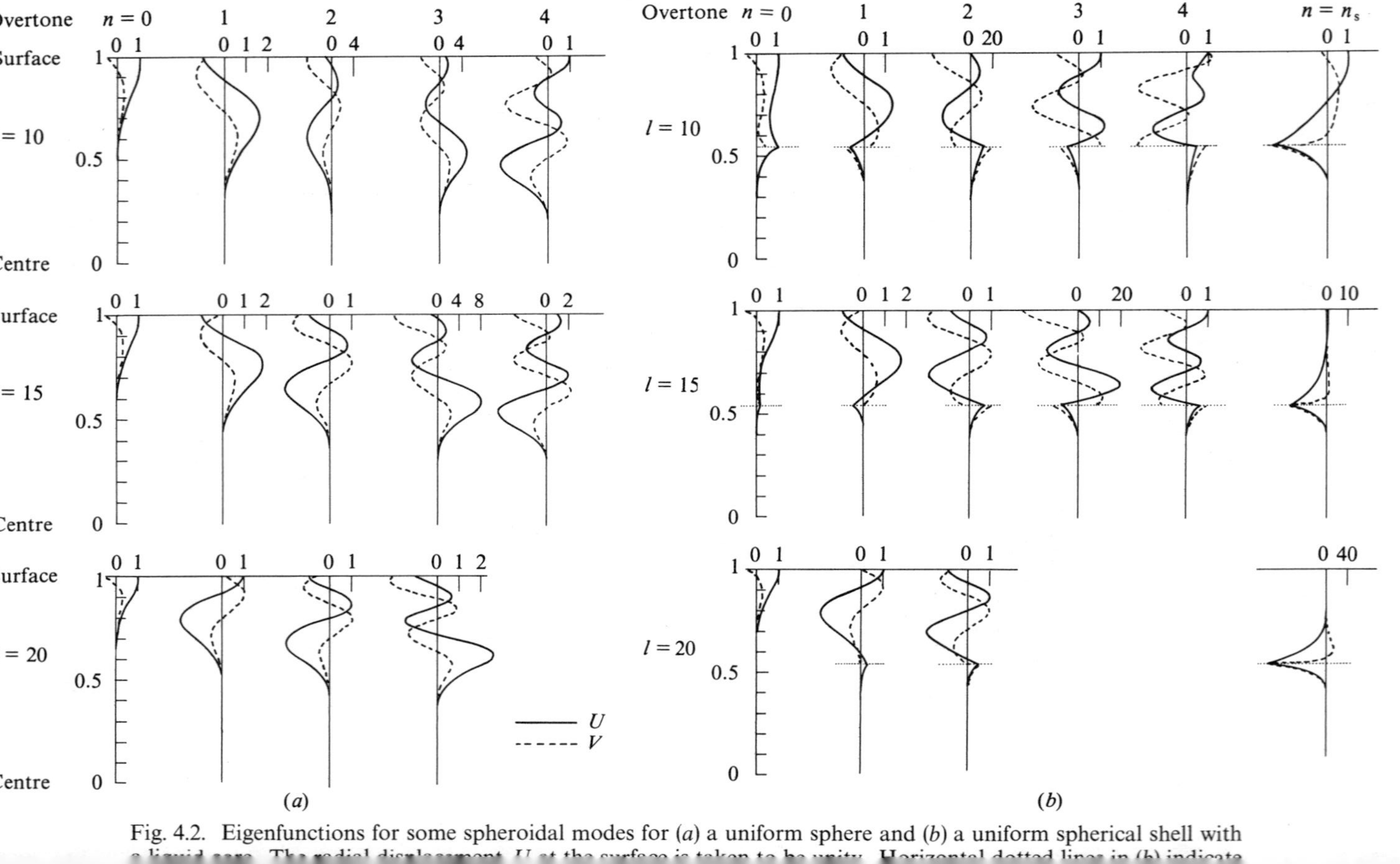

Fig. 4.2. Eigenfunctions for some spheroidal modes for (*a*) a uniform sphere and (*b*) a uniform spherical shell with a liquid core. The radial displacement U at the surface is taken to be unity. Horizontal dotted lines in (*b*) indicate

quency $\eta_o = k_o a$ as function of Legendre degree l for several low order radial overtones n (dotted lines). For comparison, curves for the same modes of a uniform elastic sphere are shown by solid lines. As would be expected, as l increases, curves for the two spheres approach each other.

A remarkable difference between the two spheres appears in an extra overtone branch (marked by overtone symbol $n = n_s$) for that with a perfect-liquid core. The physical character of the modes belonging the overtone branch $n = n_s$ is seen clearly in Fig. 4.2(*b*) which shows eigenfunctions for modes $l = 10, 15, 20, n = 0, 1, 2, 3, 4, n_s$. Corresponding modes for a uniform sphere are shown in Fig. 4.2(*a*) for comparison. In each mode, the radial displacement at the surface is standardised to unity. Solid and broken curves represent radial and transverse (latitudinal or longitudinal) displacements, respectively. The horizontal dotted lines near $r/a = 0.54$ in the right-hand figures show the position of the core–mantle boundary.

From the figure, we can observe that:

(i) Eigenfunctions for the two systems show quite similar features in the upper mantle for all corresponding modes. But there are naturally differences due to core motions, and these influence the eigenfunctions in the lower mantle.

(ii) For a uniform spherical shell with a liquid core, the transverse displacement shows discontinuity at the core–mantle boundary, a natural consequence of discarding the condition of continuity of transverse displacement on the core boundary.

(iii) Modes belonging to the special branch $n = n_s$ show large amplitudes at the core boundary. The amplitude at the core boundary relative to that on the surface increases rapidly as l increases. Larger amplitudes are concentrated in a thinner layer about the core boundary as the Legendre degree l increases. These features suggest that the modes considered are the interface waves first described by Stoneley (1924).

The expressions for phase- and group-velocities (2.5.8) and (2.5.10), hold for waves travelling on the surface. Each mode of free oscillation can be considered as a standing wave composed of equal and opposite surface waves travelling along a meridian on the surface. This

Table 4.1.

Medium	S-wave velocity	Density	P-wave velocity
1. Solid	c_{So}	ρ	$c_{Po} = \sqrt{3}c_{So}$
2. Liquid	$c_{Si} = 0$	$\rho^* = 2.2\rho$	$c^* = 1.22c_{So}$

character is deduced from the functional form of displacement in terms of ϑ alone, independent of r and φ. Therefore the phase-velocity (c) and group-velocity (c_{gS}) of the mode with $n = n_s$, assuming that they travel along the *interface*, are given by the dimensionless expressions

$$c/c_{So} \approx c_{gS}/c_{So} \approx 1.125 \times 3470/6370 \approx 0.613. \qquad (4.2.3)^\dagger$$

(The factor 1.125 for c_{gS}/c_{So} is obtained graphically, using Fig. 4.1.) The velocity of the Stoneley wave along a plane interface between media having the constants shown in Table 4.1 can be read from a diagram of a paper by Satô (1954), and is confirmed to be about 0.6 c_{So} Thus it seems plausible that the modes on the peculiar branch $n = n_s$ belong to interface waves along the core–mantle boundary.

4.3 Effect of finite rigidity of the core

The effect of finite rigidity of the whole core is difficult to find analytically, so in this section we try a numerical approach in order to obtain some understanding of the change of pattern of modes when the core changes from solid of finite rigidity to perfect-liquid.

In this chapter, the core has been assumed up to now to be a perfect-liquid, though whether the core is perfect-liquid or has finite rigidity has long been an important geophysical problem. Jeffreys (1976, § 3.07) finds no convincing evidence for a non-zero value of

†The velocity of the Stoneley wave is not a function of frequency ω, that is, the phase velocity $c = \omega/\xi$ is constant. Therefore, from the definition of group velocity, we have $c_{gS} = d\omega/d\xi = d(\xi c)/d\xi = c$, that is, the group- and phase-velocities take the same value.

core rigidity; nevertheless, it is worth while to see what would be expected in an Earth with a core of small but non-zero rigidity.

A model consisting of a uniform mantle and a uniform core exhibited a gravest period of free toroidal oscillation of 42.6 min for a perfect-liquid core and 33.8 min for a rigid core. Satô & Matumoto (1961) discovered that the fundamental toroidal free period for a sphere with a core of finite rigidity lay not between but outside the two values cited above, and the period increased indefinitely when the rigidity of the core tended to zero. In order to give a satisfactory explanation of this apparently strange phenomenon, an intensive study was undertaken in Japan.

Spheroidal oscillation The toroidal oscillation is clearly simpler to attack, but since more extensive work has been published on the spheroidal oscillation (Usami, Kotake & Satô, 1967), we deal with that first.

We use the model defined by (4.2.2). Note that this model gives $\lambda_i/\mu_o = 4.95$; λ_i is kept constant throughout this section. The rigidity of the core is introduced through R, the ratio of S-wave velocities in core and mantle, thus $R = c_{Si}/c_{So}$.

The non-dimensional frequency $\eta_o(=k_o a)$ calculated from (4.1.8) is shown as a function of l in Fig. 4.3. The thick solid line shows frequencies for a uniform sphere. The other three lines are for models consisting of uniform mantle and core. The thin solid line, dotted line and chain line are for systems in which R takes the values 0.6, 0.2 and 0.0, respectively ($R = 0.0$ refers to the liquid core). The overtone number is given for the solid sphere (thick solid line) and for a liquid core (chain line). The figure shows that:

(i) For small values of R there appear many branches of overtone mode (dotted lines). The number of such modes below the fundamental mode for a uniform sphere (thick solid line) increases as l increases. Some parts of such modes (dotted line) nearly coincide with the modes for a perfect-liquid core (chain line).

(ii) For a larger value of $R(=0.6)$, overtone branches such as those described above do not occur.

Physically speaking, when l tends to infinity, the effect of the core

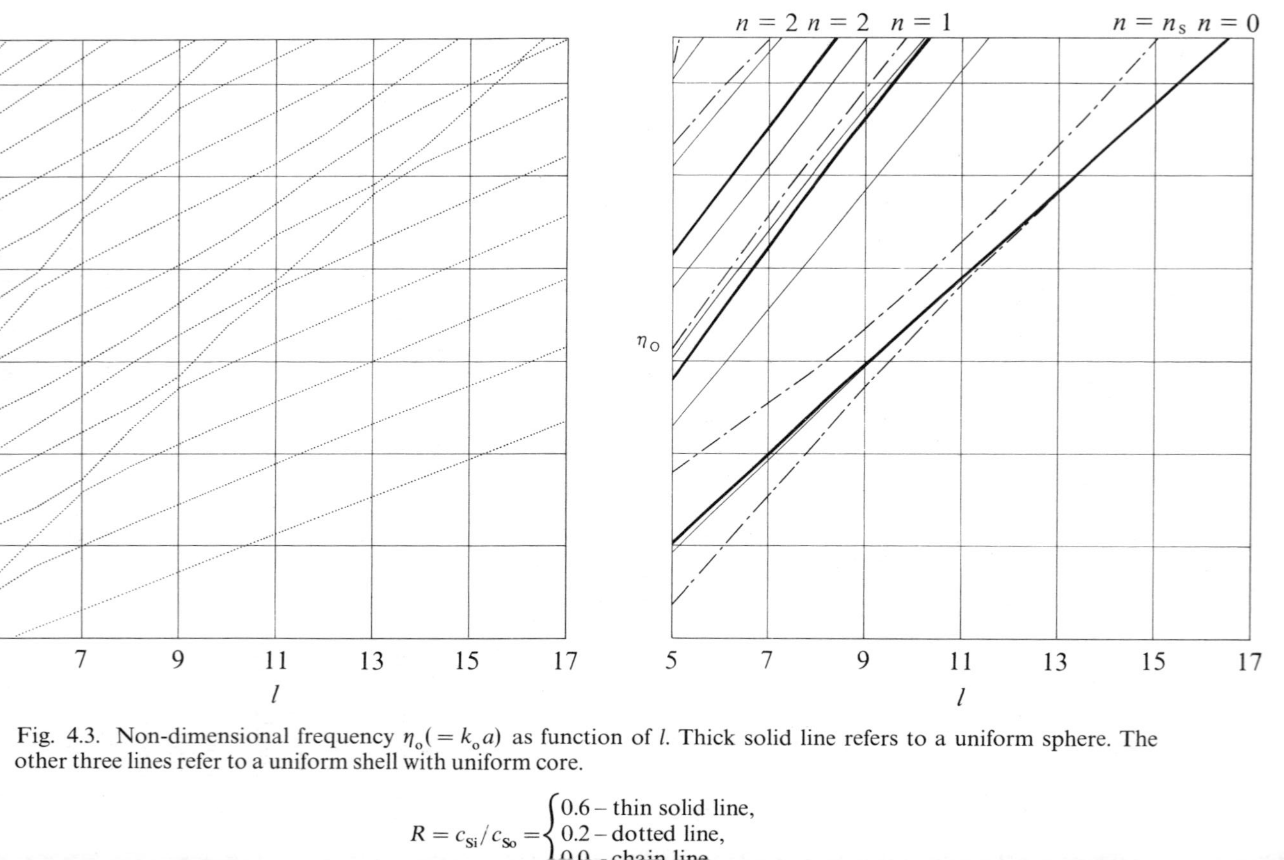

Fig. 4.3. Non-dimensional frequency $\eta_o(= k_o a)$ as function of l. Thick solid line refers to a uniform sphere. The other three lines refer to a uniform shell with uniform core.

$$R = c_{Si}/c_{So} = \begin{cases} 0.6 - \text{thin solid line,} \\ 0.2 - \text{dotted line,} \\ 0.0 - \text{chain line} \end{cases}$$

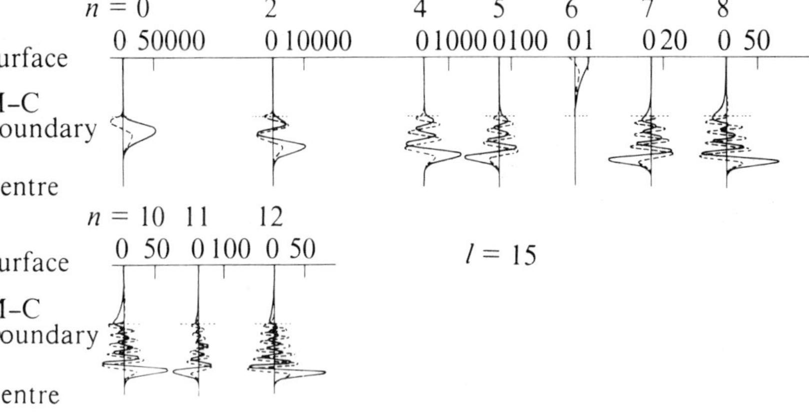

Fig. 4.4. Radial distribution of displacement for $R = 0.2$. Radial component – solid line (surface amplitude taken as unity). Transverse component – dotted line.

should become negligible and all the curves in Fig. 4.3 should approach to those for a uniform sphere. However, for $R = 0.2$, such a tendency does not seem to appear.

Fig. 4.4 shows eigenfunctions for some overtones with Legendre degree $l = 15$. It is remarkable that most of these eigenfunctions show oscillatory features in the core and are monotonic in the mantle. The amplitude of the oscillatory motion in the core is large compared with the amplitude on the surface. One exception in Fig. 4.4 is for the overtone number $n = 6$. The eigenfrequency of this mode is nearly equal to that of a sphere with a perfect-liquid core. This and other examples lead us to the conclusion that a mode having an eigenfrequency close to that of a model with perfect-liquid core has a similar eigenfunction.

The argument of the spherical Bessel function j_l which appears in the expression for $\mathbf{u}_1$ in the core is $h_i r$. In the frequency equation this becomes

$$h_i b = \frac{\omega b}{c_{Pi}} = \eta_o \frac{b}{a} \frac{c_{So}}{c_{Pi}}, \tag{4.3.1}$$

while that appearing in $\mathbf{u}_3$ is

$$k_i b = \frac{\omega b}{c_{Si}} = \eta_o \frac{b}{a} \frac{c_{So}}{c_{Si}} = \frac{\eta_o}{R} \frac{b}{a}. \tag{4.3.2}$$

When the rigidity of the core is small, that is, when the value R is small, $k_i b$ becomes large and consequently, from the property of the spherical Bessel function (Fig. 2.1), the contribution of $\mathbf{u}_3$ for small r is the main cause of the large amplitudes and oscillatory features of the displacement near the centre of the core.

For an exceptional mode ($l = 15, n = 6$) in Fig. 4.4 the eigenfrequency almost coincides with that for a perfect-liquid core. So for this mode the core with small rigidity behaves like a liquid core. In other words, the coefficient of $\mathbf{u}_3$ in the core obtained from (4.1.8) is expected to have a value almost zero.

When the contrast of impedance between two media in contact is large, most of the energy of a wave incident on the boundary is reflected and only a small part of the energy is transmitted to the other medium (Brekhovskikh, 1960 §§ 3.1–3.4). Thus, the feature shown in Fig. 4.4 where the core oscillates with larger amplitudes than those in the mantle is to be attributed to the large impedance contrast corresponding to $R = 0.2$. In other words, such a mode, in which most of the energy is trapped in the core, can be understood as an oscillation of a core surrounded by comparatively rigid wall.

Let us consider the behaviour of the non-dimensional frequency η_o as a function of R. Fig. 4.5 shows an example in which the Legendre degree l is fixed at 10. Fig. 4.5(*a*) shows the non-dimensional frequency η_o. Each horizontal dotted line indicates the non-dimensional frequency when the core is liquid. Part of Fig. 4.5(*a*) is given in Fig. 4.5(*c*) on a magnified scale. Fig. 4.5(*b*) shows the relative spectral amplitudes of the modes, solid and dashed curves referring to the radial and tangential components, respectively. The lines at the left and right sides of Figure (*b*) show the limiting values of spectral amplitudes for the radial component (heavy line) and tangential component (dotted line) when R tends to 0 and 1 respectively. They refer to the mode ($l = 10, n = 0$). Dots on the curve for $n = 12$ (Figure (*c*)) indicate modes of which eigenfunctions are given in Fig. 4.6. Fig. 4.5 shows that spectral amplitudes of modes having non-dimensional frequency η_o close to that for the sphere with liquid core are large and that the spectral amplitude diminishes rapidly as η_o departs from that for a liquid-core model. In other words, only modes having η_o close to that for a liquid-core model are observable.

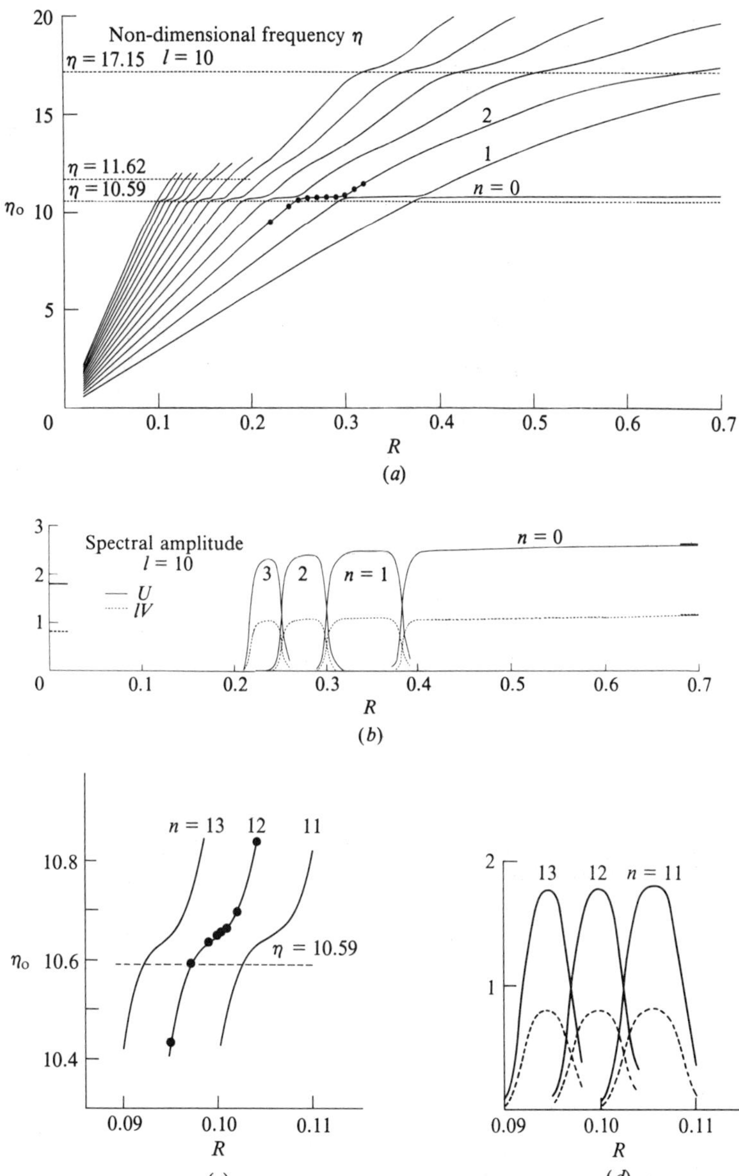

Fig. 4.5. (*a*) Non-dimensional frequency $\eta_o (= k_o a)$ as a function of R for $l = 10$. Horizontal dotted lines indicate values of η_o for $R = 0$. (*b*) Corresponding spectral amplitudes. Radial component – solid line, transverse component – dotted line. (*c*), (*d*) Detail in neighbourhood of $R = 0.1$.

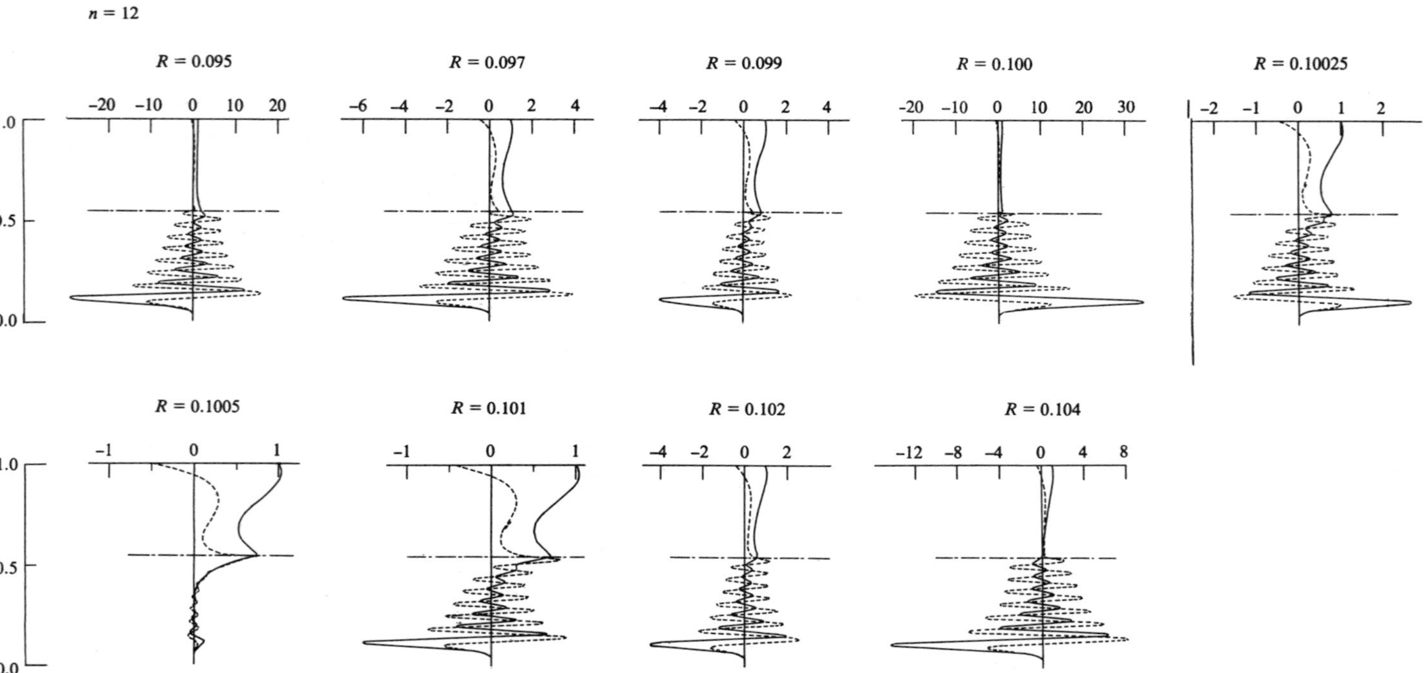

Fig. 4.6. Eigenfunctions of modes corresponding to points marked in Fig. 4.5(*c*). Radial component – solid line (surface amplitude taken as unity), transverse component – dotted line. Horizontal chain line indicates core–mantle boundary.

Fig. 4.6 suggests the possibility that special values of R, that is, special values of the core rigidity μ_c, make the oscillatory feature and large amplitudes in the core disappear.

We see from Fig. 4.5(*a*) that such a special value of core rigidity μ_c, and corresponding non-dimensional frequency close to that for a liquid-core model, are not found for large values of R. From Fig. 4.5(*b*), the spectral amplitude becomes a maximum near μ_c. As R decreases, the spectral maximum becomes sharper and special values μ_c appear at shorter intervals. When the rigidity of the core is non-zero and small, the frequency corresponding to μ_c is nearly equal to, but slightly larger than, that of the liquid core.

The fact that modes having eigenfrequencies close to those for a perfect-liquid core show predominance is illustrated in Fig. 4.7. This figure shows spectral amplitudes of several modes for two values of R, namely, 0.0 and 0.2. The abscissa represents non-dimensional frequency η_0. The numeral beside a spectral line indicates overtone number n. (The absence of an overtone number means that the corresponding mode has negligibly small spectral amplitude.) The figure shows that modes with eigenfrequencies close to those for a liquid-core model ($R = 0.0$) predominate, their spectral amplitudes being nearly equal to those of the corresponding modes for a liquid-core model. When the figure is examined carefully, it is seen that, for $R = 0.2$, corresponding to a certain mode of the liquid-core model there appear two modes of comparable amplitude. (Examples: $R = 0.2, l = 5, n = 4$ and 5 correspond to a mode of $R = 0.0, l = 5, n = 1$ for radial component u, and $R = 0.2$, $l = 10, n = 11$ and 12 correspond to a mode of $R = 0.0, l = 10, n = 2$ for both radial and tangential components.) In such cases, in place of one spectral line for the liquid-core model, it is possible to find two adjacent spectral lines of comparable order near the eigenfrequency for a liquid-core model. This phenomenon has been named '*soft core spectral splitting*' (Satô, 1964), a splitting physically different from those due to rotation of the Earth (Chapter 7) and ellipticity of the Earth (Chapter 10). This phenomenon was first recognised for toroidal oscillation of an elastic sphere with a uniform mantle and a core of finite rigidity (ibid.).

It is conjectured that *soft core spectral splitting*, detected in the pattern of distribution and amplitude of observable spectra, may give

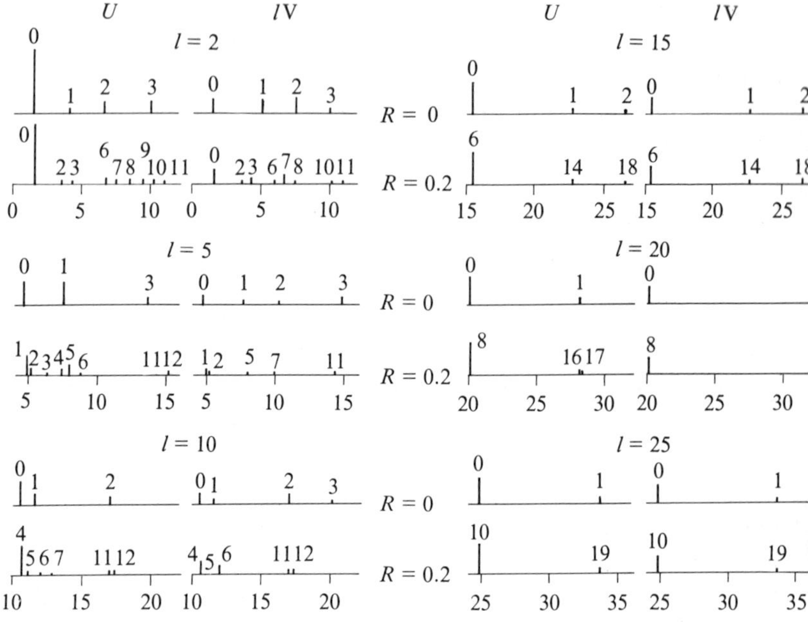

Fig. 4.7. Spectral amplitudes for some spheroidal modes of a uniform shell with uniform core, as functions of non-dimensional frequency η_0. The number beside a spectral line indicates the radial overtone number n. (The absence of an overtone number indicates that the mode has negligibly small spectral amplitude.)

a method for the determination of the rigidity of the Earth's outer core, since that pattern is sensitive to changes in R.

The discussion in this section may be summarised as follows. From our basic study of modal solutions for spheroidal free oscillations of an elastic sphere with a uniform mantle and a uniform core of finite rigidity, the fundamental characteristics of each mode are obtained. These characteristics are compared in Table 4.2 with those of a sphere with a perfect-liquid core. Here the question emerges: How can these mutually contradictory characteristics be related as the rigidity of the core tends to zero? The answers given in this section are:

(i) There are transition phenomena from one radial overtone to the next for a model of finite core rigidity, as seen in Fig. 4.3. Frequencies of these lie near the non-dimensional frequency

Table 4.2. *Comparison of characteristics of modal solutions for spheroidal free oscillations between uniform elastic sphere with* (*a*) *liquid core and* (*b*) *uniform core of finite rigidity.*

	(*a*)	(*b*)
Model	Uniform mantle with a perfect-liquid core	Uniform mantle with a uniform core of finite rigidity
Displacements at the core boundary	Horizontal displacements are discontinuous across the core boundary	Horizontal displacements (in mantle and core) are continuous across the core boundary
Special overtone branch	Except for a special branch corresponding to the Stoneley–wave along the core–mantle boundary (§ 4.2), there are no special radial overtones	Many more higher radial overtone branches appear as the rigidity of the core decreases
Displacement distribution in the core	The displacement is larger in the mantle than in the core, where the motion diminishes exponentially as the depth increases	The radial distribution of displacement in the core shows oscillatory features and has large amplitudes compared with those in the mantle

curve of the liquid core model. Modes on the transition segments (i.e. modes having eigenfrequencies close to those of the liquid-core model) show large spectral amplitudes.

(ii) In the graph of frequency against rigidity, there is found a special core rigidity μ_c on each overtone branch which displays an eigenfunction quite similar to that for a liquid core. The spectral amplitude shows its maximum near μ_c; as R decreases the maximum becomes sharper and values of μ_c appear at shorter intervals. The eigenfrequency corresponding to any μ_c is close to that for a liquid-core model.

(iii) As R tends to zero, modes showing features contradictory to those of a liquid-core model can be neglected as far as the contribution to the surface displacement is concerned. Thus, it is to be expected that modes showing characteristics similar to those of modes of liquid-core model remain.

Toroidal oscillations For toroidal oscillations the following features have been described (Satô, 1964):

(i) When the core has small finite rigidity, there appear many more spectral lines than for a liquid core. Among these, modes which have frequencies nearly equal to those of the liquid-core model predominate.

(ii) Some modes show *soft core spectral splitting* according to the value of R.

(iii) Motion in the core exhibits oscillatory features and amplitudes which are large compared with those in the mantle.

From these characteristics it is to be expected that the transition from a core of finite rigidity to a perfect-liquid core in toroidal oscillations will be found to be similar to the transition for spheroidal oscillations.

5

OSCILLATION OF A SELF-GRAVITATING LAYERED SPHERE

5.1 Derivation of equations of motion

So far we have studied the problem of free oscillation of a sphere for simple models only – a uniform sphere and a two-layered sphere consisting of uniform layers. We have neglected the effects of gravity, heterogeneity, rotation, ellipticity and anelasticity. Simple models are effective for the consideration of fundamental properties of the free oscillations and of the meaning of each mode, but they are far from the complexity of the Earth. We will now introduce the main further characteristics of the Earth one by one: gravity in this chapter, radial heterogeneity in Chapter 6, and rotation in Chapter 7. Complexities which go beyond the scope of this book will be treated briefly in Chapter 10.

The equations of small motion of a perfectly elastic, heterogeneous and isotropic medium take, in polar coordinates, the form (Love, 1927, § 59) (see also Appendix § A.2)

$$\partial_r \begin{Bmatrix} \widehat{rr} \\ \widehat{\vartheta r} \\ \widehat{\varphi r} \end{Bmatrix} + \frac{1}{r}\partial_\vartheta \begin{Bmatrix} \widehat{r\vartheta} \\ \widehat{\vartheta\vartheta} \\ \widehat{\varphi\vartheta} \end{Bmatrix} + \frac{1}{r\sin\vartheta}\partial_\varphi \begin{Bmatrix} \widehat{r\varphi} \\ \widehat{\vartheta\varphi} \\ \widehat{\varphi\varphi} \end{Bmatrix}$$

$$+\frac{1}{r}\begin{Bmatrix} 2\widehat{rr} - \widehat{\vartheta\vartheta} - \widehat{\varphi\varphi} + \widehat{r\vartheta}\cot\vartheta \\ (\widehat{\vartheta\vartheta} - \widehat{\varphi\varphi})\cot\vartheta + 3\widehat{r\vartheta} \\ 3\widehat{r\varphi} + 2\widehat{\vartheta\varphi}\cot\vartheta \end{Bmatrix} + \begin{Bmatrix} F_r \\ F_\vartheta \\ F_\varphi \end{Bmatrix} = \rho\partial_t^2 \begin{Bmatrix} u \\ v \\ w \end{Bmatrix}, \tag{5.1.1}$$

where $\widehat{rr}, \widehat{r\vartheta}, \ldots$, are the stress components, $F_r, F_\vartheta, F_\varphi$ the components of body force **F** and u, v, w the displacement components in the directions of r, ϑ, φ increasing. These equations are the equations of motion in the radial, latitudinal and longitudinal directions respectively. The fourth term on the left-hand side of (5.1.1) arises from the curvature of the polar coordinate lines.

We introduce the effect of gravity following the procedure of Love (1911). The density ρ in the equilibrium state and Lamé's constants λ and μ are assumed to be functions of r only. In a self-gravitating system an elastic sphere is not free from stress even in the equilibrium state. We assume that a reference state exists in which the sphere is initially in hydrostatic equilibrium. In such an equilibrium state, we have, by consideration of radial forces on an element,†

$$-\rho_0 g_0 = \partial_r p_0, \tag{5.1.2}$$

where g is the gravitational acceleration, p the hydrostatic pressure (which are functions of r only), and the suffix 0 labels the quantity in the equilibrium state. The gravitational potential ψ_0 satisfies Poisson's equation

$$\nabla^2 \psi_0 = -4\pi G \rho_0, \tag{5.1.3}$$

where G is the universal constant of gravitation. The gravitational acceleration g_0 is obtained from ψ_0 by the relation

$$\partial_r \psi_0 = -g_0. \tag{5.1.4}$$

In the disturbed state the density and gravitational potential may be expressed as

$$\rho = \rho_0 + \rho', \qquad \psi = \psi_0 + \psi', \tag{5.1.5}$$

where ρ' and ψ' are terms due to disturbance and are assumed to be first-order small quantities.

The principle of conservation of mass states that the net amount of mass that moves into any volume V through the fixed surface S enclosing V in the displacement must be equal to the net change of mass in V. That is

$$\int_S \rho u_n \, \mathrm{d}S = \delta \int \rho \, \mathrm{d}V, \tag{5.1.6}$$

†Fundamental properties of gravity fields may be found in Ramsey (1961) and in many similar texts.

where u_n is the displacement along the normal to S drawn into V. By Gauss's theorem, (5.1.6) reduces to

$$\int_V \left\{ \delta\rho + \frac{\partial(\rho u_i)}{\partial x_i} \right\} \mathrm{d}V = 0. \tag{5.1.7}$$

Since V is arbitrary, we have

$$\begin{aligned} \rho' = \delta\rho_0 &= -\operatorname{div}(\rho_0 \mathbf{u}) \\ &= -\rho_0 \Delta - u\partial_r \rho_0, \end{aligned} \tag{5.1.8}$$

Δ being dilatation as before. Poisson's equation is also true for the disturbed state. Hence the gravitational potential ψ' satisfies the equation

$$\nabla^2 \psi' = 4\pi G(\rho_0 \Delta + u\partial_r \rho_0). \tag{5.1.9}$$

In any slightly deformed state, the stress is assumed to consist of the sum of the initial stress and the stress due to the small displacement. The latter is expressed by (2.3.3) and (2.3.4). The former is not the original hydrostatic pressure at the (geometrical) point now under consideration, but the hydrostatic pressure which obtained initially at the *material point* under consideration. This was the original hydrostatic pressure at the geometrical point $(r-u, \vartheta, \varphi)$, namely

$$-(p_0 - u\partial_r p_0).$$

Hence the stress components in equation (5.1.1) are given by

$$\left.\begin{aligned} \widehat{rr} &= -(p_0 - u\partial_r p_0) + \lambda\Delta + 2\mu e_{rr}, \\ \widehat{\vartheta\vartheta} &= -(p_0 - u\partial_r p_0) + \lambda\Delta + 2\mu e_{\vartheta\vartheta}, \\ \widehat{\varphi\varphi} &= -(p_0 - u\partial_r p_0) + \lambda\Delta + 2\mu e_{\varphi\varphi}, \\ \widehat{r\vartheta} &= 2\mu e_{r\vartheta}, \\ \widehat{\vartheta\varphi} &= 2\mu e_{\vartheta\varphi}, \\ \widehat{\varphi r} &= 2\mu e_{\varphi r}. \end{aligned}\right\} \tag{5.1.10}$$

The effect of gravity appears also in the body-force term. The body-force $\mathbf{F}$ $(F_r, F_\vartheta, F_\varphi)$ derived from gravity is

$$\mathbf{F} = \rho \operatorname{grad} \psi. \tag{5.1.11}$$

Since quantities with suffix 0 are known when a model sphere is defined and ρ' is found from ρ_0 and (u, v, w) by (5.1.8), the equations of motion (5.1.1) are three simultaneous equations in the unknowns u, v, w and ψ'. Thus we need one more equation, and that is Poisson's equation (5.1.9). In other words, the field equations for a gravitating elastic sphere are the set of four simultaneous equations consisting of (5.1.1) and (5.1.9). Eliminating stresses and strains in favour of the dependent variables u, v, w, ψ', and neglecting small quantities of the second order, we obtain

$$\left.\begin{aligned}
&\partial_r(\lambda\Delta) + 2\partial_r(\mu\partial_r u) + \frac{\mu}{r}\left(4\partial_r u - 4\frac{u}{r} + \frac{\cot\vartheta}{r}\partial_\vartheta u + \frac{1}{r}\partial_\vartheta^2 u\right.\\
&\qquad \left. + \frac{1}{r\sin^2\vartheta}\partial_\varphi^2 u\right) + \frac{\mu}{r}\left(-\frac{3\cot\vartheta}{r}v + \cot\vartheta\,\partial_r v\right.\\
&\qquad \left. + \partial_r\partial_\vartheta v - \frac{3}{r}\partial_\vartheta v\right) + \frac{\mu}{r\sin\vartheta}\left(-\frac{3}{r}\partial_\varphi w + \partial_r\partial_\varphi w\right)\\
&\qquad + \rho_0\partial_r\psi' - \rho_0\partial_r(ug_0) + \rho_0 g_0\Delta = \rho_0\partial_t^2 u,\\
&\frac{1}{r}\partial_\vartheta(\lambda\Delta) + \partial_r\left\{\mu\left(\partial_r v - \frac{v}{r} + \frac{1}{r}\partial_\vartheta u\right)\right\} + \frac{5\mu}{r^2}\partial_\vartheta u\\
&\qquad + \frac{\mu}{r}\left(-2\frac{v}{r}\cot^2\vartheta - 3\frac{v}{r} + 3\partial_r v + \frac{2}{r}\cot\vartheta\partial_\vartheta v\right.\\
&\qquad \left. + \frac{1}{r\sin^2\vartheta}\partial_\varphi^2 v + \frac{2}{r}\partial_\vartheta^2 v\right)\\
&\qquad + \frac{\mu}{r\sin\vartheta}\left(\frac{1}{r}\partial_\varphi\partial_\vartheta w - \frac{3}{r}\cot\vartheta\partial_\varphi w\right)\\
&\qquad + \frac{\rho_0}{r}\partial_\vartheta\psi' - \frac{1}{r}\rho_0 g_0\partial_\vartheta u = \rho_0\partial_t^2 v,\\
&\frac{1}{r\sin\vartheta}\partial_\varphi(\lambda\Delta) + \partial_r\left\{\mu\left(\frac{1}{r\sin\vartheta}\partial_\varphi u + \partial_r w - \frac{w}{r}\right)\right\}\\
&\quad + \frac{5\mu}{r^2\sin\vartheta}\partial_\varphi u + \frac{\mu}{r}\left\{\frac{4\cot\vartheta}{r\sin\vartheta}\partial_\varphi v + \partial_\vartheta\left(\frac{1}{r\sin\vartheta}\partial_\varphi v\right)\right\}\\
&\quad + \frac{\mu}{r}\left\{-\frac{3}{r}w - \frac{2}{r}\cot^2\vartheta w + 3\partial_r w + \frac{2}{r}\cot\vartheta\partial_\vartheta w\right.
\end{aligned}\right\} \quad (5.1.12)$$

$$+\frac{2}{r\sin^2\vartheta}\partial_\varphi^2 w+\frac{1}{r}\partial_\vartheta^2 w-\frac{1}{r}\partial_\vartheta(w\cot\vartheta)\Bigg\}$$

$$+\frac{\rho_0}{r\sin\vartheta}\partial_\varphi\psi'-\frac{\rho_0\vartheta_0}{r\sin\vartheta}\partial_\varphi u=\rho_0\partial_t^2 w,$$

$$\partial_r^2\psi'+\frac{2}{r}\partial_r\psi'+\frac{1}{r^2}\partial_\vartheta^2\psi'+\frac{\cot\vartheta}{r^2}\partial_\vartheta\psi'$$

$$+\frac{1}{r^2\sin^2\vartheta}\partial_\varphi^2\psi'=4\pi G(\rho_0\Delta+u\partial_r\rho_0).$$

On the left-hand side of the first three equations of (5.1.12), the leading terms come from grad $(\lambda\Delta)$, forming the force due to dilatation. These, with the succeeding terms which contain u, v and w, represent the gradient of that part of the stress which is due to the strain. The last two or three terms show the effect of gravity; among these are the gradient of the additional potential ψ' and the terms contributed by the initial hydrostatic stress.

We now redefine the toroidal oscillations as those in which the radial displacement and dilatation are both zero. Then by (5.1.8) the density change ρ' is also zero, and thus the additional gravitational potential ψ' is everywhere zero. Moreover, the force $u\partial_r p_0$ due to the change of hydrostatic pressure is zero. Thus of the terms introduced by the effect of gravity there remain only

$$\operatorname{grad}(-p_0)+\rho_0\operatorname{grad}\psi_0\,;$$

this expression is zero from (5.1.2) and (5.1.4). Thus it is proved that the toroidal oscillations of an elastic sphere are unaffected by self-gravitation.

The boundary conditions on a surface of discontinuity are of two kinds, elastic and gravitational. The former demand, as before, continuity of displacement and stress components at the surface of discontinuity; these are expressed as

$$\left.\begin{array}{lll} u_1=u_2, & v_1=v_2, & w_1=w_2, \\ \widehat{rr}_1=\widehat{rr}_2, & \widehat{r\vartheta}_1=\widehat{r\vartheta}_2, & \widehat{r\varphi}_1=\widehat{r\varphi}_2, \end{array}\right\} \qquad (5.1.13)$$

where suffixes 1 and 2 refer to the media on the two sides of the discontinuity. The latter demand the continuity of gravitational potential and its normal gradient at the boundary surface (Appendix

§ A.3), namely:

$$\psi_1 = \psi_2, \qquad \partial_n \psi_1 = \partial_n \psi_2, \tag{5.1.14}$$

where ∂_n denotes the derivative normal to the boundary surface. The gravitational potential ψ consists of ψ_0 (equilibrium potential) with ψ' (additional potential due to disturbance). ψ_0 satisfies the condition (5.1.14). Hence it is evident that

$$\psi'_1 = \psi'_2 \tag{5.1.15}$$

at the boundary surface. As for the gradient, its value on a deformed surface reduces, by the use of the relation

$$\frac{\partial}{\partial n} = \frac{\partial r}{\partial n}\frac{\partial}{\partial r} + \frac{\partial \vartheta}{\partial n}\frac{\partial}{\partial \vartheta} + \frac{\partial \varphi}{\partial n}\frac{\partial}{\partial \varphi},$$

to

$$\begin{aligned} \partial_n \psi &= (\partial_n \psi)_{\mathrm{a}} + u(\partial_r \partial_n \psi)_{\mathrm{a}} \\ &= \{\partial_n r \partial_r (\psi_0 + \psi')\}_{\mathrm{a}} + \{\partial_n \vartheta \partial_\vartheta (\psi_0 + \psi')\}_{\mathrm{a}} + \{\partial_n \varphi \partial_\varphi (\psi_0 + \psi')\}_{\mathrm{a}} \\ &\quad + u[\partial_r \{\partial_n r \partial_r (\psi_0 + \psi') + \partial_n \vartheta \partial_\vartheta (\psi_0 + \psi') + \partial_n \varphi \partial_\varphi (\psi_0 + \psi')\}]_{\mathrm{a}}. \end{aligned} \tag{5.1.16}$$

The suffix a indicates that the term must be evaluated at the undisturbed surface.

We are considering small disturbances, and so we will assume that $\partial_n \vartheta, \partial_n \varphi$ are first-order small quantities, and that $\partial_n r$ differs from unity by a second-order small quantity. Remembering that ψ_0 is a function of r only, neglecting higher-order small terms and employing (5.1.3), (5.1.4), we have

$$\partial_n \psi \approx \partial_r \psi + u \partial_r^2 \psi_0 = -g_0 + \partial_r \psi' + u(-4\pi G \rho_0 + 2g_0/r).$$

Since g_0 and u are continuous on a boundary, (5.1.14) implies the continuity of $\partial_r \psi' - 4\pi G \rho_0 u$. Thus we have

$$(\partial_r \psi' - 4\pi G \rho_0 u)_1 = (\partial_r \psi' - 4\pi G \rho_0 u)_2 \tag{5.1.17}$$

as boundary condition. Summing up, the boundary conditions at a surface of discontinuity in a self-gravitating elastic sphere are (5.1.13), (5.1.15) and (5.1.17).

5.2 Effect of gravity on the spheroidal oscillation of a uniform sphere

In this section, we assume that the sphere is uniform; then the parameters λ, μ and ρ_0 are constant and the gravitational acceleration g_0 is given by

$$g_0 = \tfrac{4}{3}\pi G\rho_0 r = \Gamma r, \tag{5.2.1}$$

where

$$\Gamma = \tfrac{4}{3}\pi G\rho_0 .$$

The equations of motion (5.1.1) now become simpler, but they are still too complicated to be capable of analytic solutions. Remembering that the analytic solution took the form (2.2.28–30) for a non-gravitational uniform elastic sphere, and considering that in this section we still treat a spherical Earth in which g_0 is a function of r only, it seems reasonable to try an elementary solution (containing surface harmonics of azimuthal order m and degree l in the same form as before):

$$\left.\begin{aligned} u &\propto U_l(r)P_l^m(\cos\vartheta)\begin{smallmatrix}\cos\\ \sin\end{smallmatrix} m\varphi, \\ v &\propto V_l(r)\partial_\vartheta P_l^m(\cos\vartheta)\begin{smallmatrix}\cos\\ \sin\end{smallmatrix} m\varphi, \\ w &\propto mV_l(r)\frac{1}{\sin\vartheta}P_l^m(\cos\vartheta)\begin{smallmatrix}(-\sin)\\ \cos\end{smallmatrix} m\varphi. \end{aligned}\right\} \tag{5.2.2}$$

The function ψ' must satisfy the last equation of (5.1.12): the dilatation on the right-hand side is obtainable from (A.1.3), which under the assumption of (5.2.2) has the form

$$\text{const} \times (\text{function of } r) \times P_l^m(\cos\vartheta)\begin{smallmatrix}\cos\\ \sin\end{smallmatrix} m\varphi .$$

Hence we assume the form

$$\psi' \propto Y_l(r)P_l^m(\cos\vartheta)\begin{smallmatrix}\cos\\ \sin\end{smallmatrix} m\varphi . \tag{5.2.3}$$

Putting (5.2.2) and (5.2.3) into (5.1.12) and seeking a solution with time variation exp $i\omega t$, we find that the second and third equations reduce to a single equation. Thus (5.1.1) give rise to three simultaneous equations in the unknowns U_l, V_l and Y_l. They are

$$\begin{aligned} &\partial_r(\lambda\Delta_l + 2\mu\dot U_l) + \frac{\mu}{r^2}\{4r\dot U_l - 4U_l + l(l+1)(3V_l - U_l - r\dot V_l)\} \\ &\qquad + \rho_0\dot Y_l + \rho_0 g_0\Delta_l - \rho_0\partial_r(g_0U_l) + \rho_0\omega^2U_l = 0, \end{aligned}$$

$$\left.\begin{aligned}\partial_r\left\{\mu\left(\dot V_l-\frac{V_l-U_l}{r}\right)\right\}+\frac{\mu}{r^2}\{5U_l-V_l-2l(l+1)V_l+3r\dot V_l\}&\\+\frac{\lambda}{r}\Delta_l+\frac{\rho_0}{r}Y_l-\frac{\rho_0 g_0}{r}U_l+\rho_0\omega^2V_l&=0,\\ \ddot Y_l+\frac{2}{r}\dot Y_l-\frac{l(l+1)}{r^2}Y_l-4\pi G(\dot\rho_0U_l+\rho_0\Delta_l)&=0,\end{aligned}\right\} \tag{5.2.4}$$

where the expression Δ_l (dilatation) is

$$\Delta_l=\dot U_l+\frac{2}{r}U_l-\frac{l(l+1)}{r}V_l. \tag{5.2.5}$$

In (5.2.4) and (5.2.5), the dot represents ∂_r as before. Equations (5.2.4) are valid for any radially heterogeneous sphere, that is, for a sphere in which the physical parameters λ, μ and ρ_0 are functions of r only.

The boundary conditions to be satisfied on the deformed outer surface $r=a+u$ are those for zero surface stress,

$$\widehat{rr}=0,\qquad \widehat{r\vartheta}=0,\qquad \widehat{r\varphi}=0,$$

and the two gravitational constraints (5.1.15) and (5.1.17). A condition of zero surface stress is, for instance, $0=\widehat{rr}_{a+u}=\widehat{rr}_a+u\partial_r(\widehat{rr})$. Putting the first relation of (5.1.10) into this expression, and neglecting small quantities of higher order, we obtain

$$(\lambda+2\mu)\dot U_l+2\frac{\lambda}{r}U_l-\frac{\lambda}{r}l(l+1)V_l=0\qquad \text{at } r=a; \tag{5.2.6}$$

similarly, for $\widehat{r\vartheta}=0$ and $\widehat{r\varphi}=0$, we obtain the single equation

$$\dot V_l-\frac{1}{r}(V_l-U_l)=0\qquad \text{at } r=a. \tag{5.2.7}$$

Let the disturbance of potential of corresponding degree outside the sphere be ψ_l'. Since the density outside the sphere is zero, we obtain from (5.1.15) and (5.1.17)

$$\left.\begin{aligned}\psi'&=\psi_l'\\ \partial_r\psi'-\partial_r\psi_l'&=4\pi G\rho_0u\end{aligned}\right\}\qquad \text{at } r=a. \tag{5.2.8}$$

ψ_l' satisfies Laplace's equation $\nabla^2\psi_l'=0$ and vanishes as r tends to infinity. Thus we obtain

$$\psi_l'\propto r^{-(l+1)}$$

Therefore

$$\partial_r \psi_l' = -\frac{l+1}{r}\psi_l' = -\frac{l+1}{r}\psi',$$

and hence we obtain

$$\partial_r \psi' + \frac{(l+1)}{r}\psi' = 4\pi G\rho_0 U_l \qquad \text{at } r = a \tag{5.2.9}$$

as the boundary condition to be satisfied by the additional potential ψ'. It should be noted that the two boundary conditions relating to the potential ψ' reduce to one condition (5.2.9).

Operating by differentiation, addition and subtraction on the first two equations of (5.2.4), we obtain a fourth-order differential equation in the unknown Δ. This equation may be solved in terms of spherical Bessel functions. Using Δ thus obtained and the auxiliary function $H = \dot{V}_l + (V_l - U_l)/r$, Pekeris & Jarosch (1958) found analytical solutions of (5.2.4) for a uniform elastic sphere (Appendix § 4.4). These are:

$$\left.\begin{aligned}
\Delta_l(\zeta) &= Rz_l(p\zeta) + Sz_l(q\zeta),\\
U_l(\zeta) &= R\left[\frac{a}{p}z_{l+1}(p\zeta) - \frac{al}{p^2\zeta}\{1+(l+1)(M(p)-1\}z_l(p\zeta)\right]\\
&\quad + S\left[\frac{a}{q}z_{l+1}(q\zeta) - \frac{al}{q^2\zeta}\{1+(l+1)(M(q)-1\}z_l(q\zeta)\right]\\
&\quad + Tl\zeta^{l-1} - T'(l+1)\zeta^{-l-2},\\
V_l(\zeta) &= R\left[\frac{a}{p}(M(p)-1)z_{l+1}(p\zeta)\right.\\
&\quad \left. - \frac{a}{p^2\zeta}\{1+(l+1)(M(p)-1)\}z_l(p\zeta)\right]\\
&\quad + S\left[\frac{a}{q}(M(q)-1)z_{l+1}(q\zeta)\right.\\
&\quad \left. - \frac{a}{q^2\zeta}\{1+(l+1)(M(q)-1)\}z_l(q\zeta)\right]\\
&\quad + T\zeta^{l-1} + T'\zeta^{-l-2},\\
Y_l(\zeta) &= -\frac{3\Gamma a^2 R}{p^2}z_l(p\zeta) - \frac{3\Gamma a^2 S}{q^2}z_l(q\zeta) + (\Gamma l - \omega^2)T\zeta^l\\
&\quad - \{\Gamma(l+1) + \omega^2\}T'\zeta^{-l-1},
\end{aligned}\right\} \tag{5.2.11}$$

where

$$\zeta = \frac{r}{a}, \qquad \beta^2 = \frac{\rho_0}{\mu} = \frac{1}{c_S^2}, \qquad \alpha^2 = \frac{\rho_0}{\lambda + 2\mu} = \frac{1}{c_P^2},$$

and

$$\left.\begin{matrix} p^2 \\ q^2 \end{matrix}\right\} = \tfrac{1}{2}a^2 \Big\{ \beta^2\omega^2 + \alpha^2(\omega^2 + 4\Gamma) \pm [\{\beta^2\omega^2 - \alpha^2(\omega^2 + 4\Gamma)\}^2 + 4l(l+1)\alpha^2\beta^2\Gamma^2]^{1/2} \Big\}, \tag{5.2.12}$$

$$M(x) = 1 + \Gamma(\omega^2 - x^2/\alpha^2\beta^2)^{-1}.$$

z_l stands for a spherical Bessel function (Usami & Satô, 1966). R, S, T and T' are constant multipliers whose ratio is to be determined by the boundary conditions. In a region containing $r = 0$, the term with T' is to be omitted.

For a *uniform sphere*, from the requirement that there must be no singularity at the centre, z_l takes the form j_l, and T' must zero. The solution (5.2.11) thus has three unknown constants R, S and T, and we have three boundary conditions (5.2.6), (5.2.7) and (5.2.9). Therefore by putting (5.2.11) into these, we obtain three homogeneous linear equations for R, S and T, and the consistency condition enables us to calculate the frequencies of the spheroidal oscillation of a self-gravitating uniform Earth.

We have chosen, for the purpose of illustration, constants which characterise a radially averaged Earth:

$$\left.\begin{aligned} a &= 6370 \text{ km}, \\ c_S &= 6.667 \text{ km/s}, \qquad c_P = 11.55 \text{ km/s}, \\ \lambda &= \mu, \\ \rho_0 &= 5.52 \text{ gm/cm}^3, \\ G &= 6.67 \times 10^{-8} (\text{CGS}), \\ \Gamma &= 154.2 \times 10^{-8} (\text{CGS}). \end{aligned}\right\} \tag{5.2.13}$$

Non-dimensional frequencies $\eta = \omega a / c_S$ obtained for this model are shown in Fig. 5.1, together with those for a non-gravitating uniform elastic sphere of the same material constants. Simple physical considerations lead us to expect that since self-gravity tightens an elastic sphere, its effect must be to shorten the period of free oscillation (i.e. to increase the non-dimensional frequency).

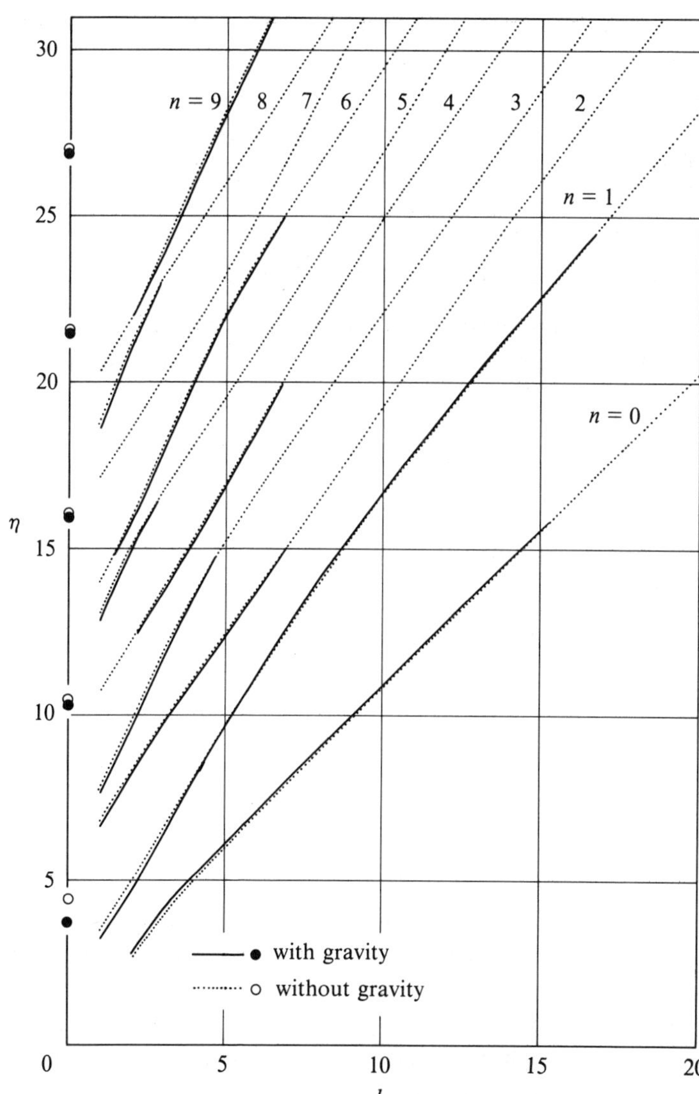

Fig. 5.1. Non-dimensional frequency η as function of l for spheroidal oscillations of a uniform elastic sphere with material constants given in (5.2.13).

This is seen in the figure to be true for the fundamental mode $n = 0$ for all l except $l = 0$. Apparently it is not true for higher overtones, suggesting a complicated effect of self-gravitation. We examine the relationship for the mode $l = 0$.

When $l = 0$, the problem reduces to the solution of

$$\partial_r^2 U + \frac{2}{r}\partial_r U - \frac{2}{r^2}U + \frac{\rho_0(\omega^2 + 4\Gamma)}{\lambda + 2\mu}U = 0 \tag{5.2.14}$$

under the condition on the surface

$$(\lambda + 2\mu)\partial_r U + 2\lambda U/r = 0 \qquad \text{at } r = a. \tag{5.2.15}$$

Denoting the non-dimensional frequency for a gravitating sphere by η_g and that for a non-gravitating sphere by η, and corresponding displacements by U_g and U, we have

$$\left.\begin{aligned} &\partial_r^2 U_g + \frac{2}{r}\partial_r U_g - \frac{2}{r^2}U_g + \frac{\rho_0(4\Gamma + \eta_g^2 c_S^2/a^2)}{\lambda + 2\mu}U_g = 0, \\ &\text{with} \\ &(\lambda + 2\mu)\partial_r U_g + 2\lambda U_g/r = 0 \qquad \text{at } r = a, \end{aligned}\right\} \tag{5.2.16}$$

and

$$\left.\begin{aligned} &\partial_r^2 U + \frac{2}{r}\partial_r U - \frac{2}{r^2}U + \frac{\rho_0 \eta^2 c_S^2/a^2}{\lambda + 2\mu}U = 0 \\ &\text{with} \\ &(\lambda + 2\mu)\partial_r U + 2\lambda U/r = 0 \qquad \text{at } r = a. \end{aligned}\right\} \tag{5.2.17}$$

These problems are identical and therefore have identical solutions, and hence

$$\eta^2 = \eta_g^2 + 4a^2\Gamma/c_S^2. \tag{5.2.18}$$

Since Γ is a positive quantity, we conclude that $\eta > \eta_g$. This means that for $l = 0$, that is, for an oscillation which is purely radial, the effect of gravity is to lengthen the eigenperiod. In our numerical example the difference of eigenfrequencies between gravitating and non-gravitating spheres is not large, and becomes smaller as n increases. This can be seen on the axis $l = 0$ in Fig. 5.1. The difference in radial eigenfunctions may also be shown to be small.

With regard to this problem, Sir Harold Jeffreys has commented: 'There was horrible trouble in Love's *Geodynamics* when he came to compressibility. He took gravity into account, and assumed constant density. This would imply that if the pressure was taken off the density would decrease towards the centre, which is badly wrong even if the composition is uniform.'

The simple example given above is vulnerable to the same criticism. But we have nevertheless included it, partly so as to record Jeffreys's criticism of Love's great classic, and partly because we do not know of any other simple published example which shows the influence of self-gravitation.

5.3 Free oscillations of a self-gravitating sphere with uniform mantle and uniform liquid core

We next consider the effect of gravity on the frequencies of spheroidal oscillations of a spherical shell with a liquid core. The mantle and the core are assumed to be uniform. Solutions of the equations of motion for both mantle and core are expressed by (5.2.10) and (5.2.11). In the mantle, the spherical Bessel function $z_l(r)$ should be a linear combination of $j_l(r)$ and $y_l(r)$. Similarly, terms containing both T and T' should be retained in the mantle. Thus, we count six independent solutions for the mantle. On the contrary, in the core, from the requirement of regularity at the centre of sphere $(r = 0)$, the spherical Bessel function must be $j_l(r)$ and the term including T' must be excluded. Thus in the core we have three independent solutions.

We must combine these nine independent solutions to satisfy three boundary conditions (5.2.6, 7, 9) and six boundary conditions on the mantle–core boundary:

$$\left.\begin{aligned}
(U_l)_o &= (U_l)_i,\\
(V_l)_o &= (V_l)_i,\\
\left[(\lambda+2\mu)\partial_r U_l + \frac{\lambda}{a}\{2U_l - l(l+1)V_l\}\right]_o &= \left[(\lambda+2\mu)\partial_r U_l \right.\\
&\qquad \left. + \frac{\lambda}{a}\{2U_l - l(l+1)V_l\}\right]_i,\\
[\mu\{\partial_r V_l - (V_l - U_l)/a\}]_o &= [\mu\{\partial_r V_l - (V_l - U_l)/a\}]_i,\\
(\psi'_l)_o &= (\psi'_l)_i,\\
(\partial_r \psi'_l - 4\pi G\rho_0 U_l)_o &= (\partial_r \psi'_l - 4\pi G\rho_0 U_l)_i,
\end{aligned}\right\} \qquad (5.3.1)$$

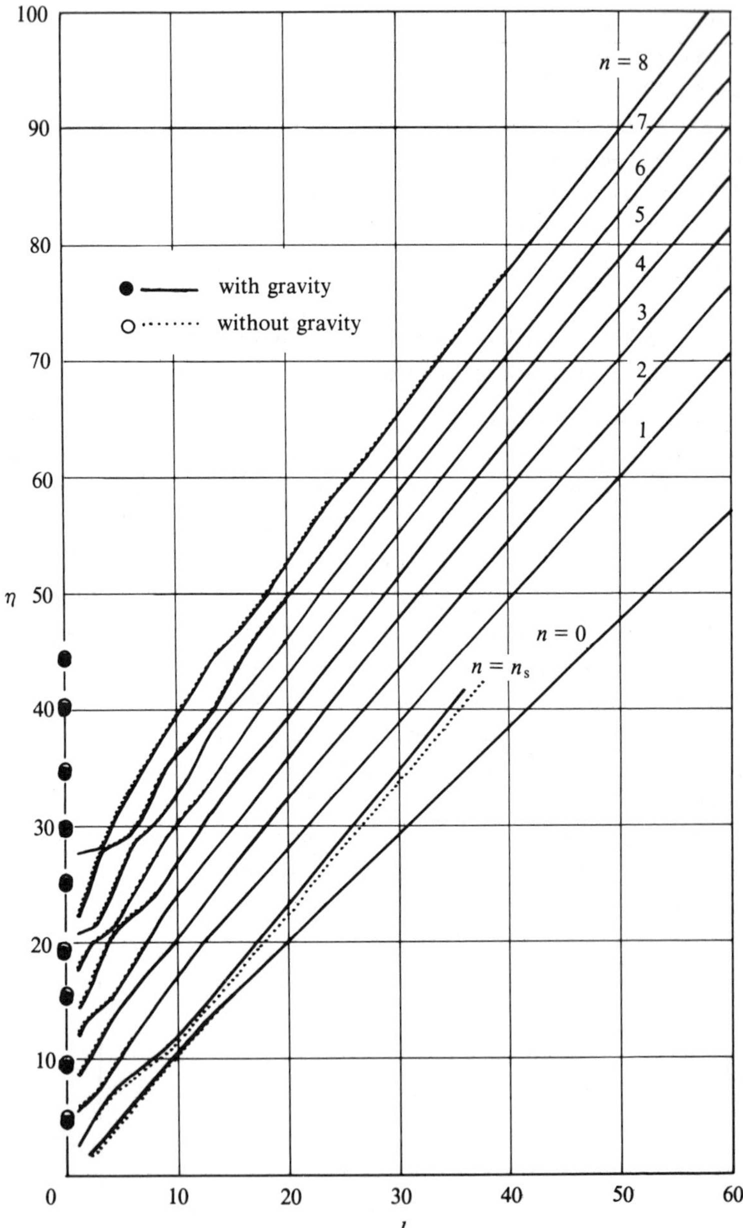

Fig. 5.2. Non-dimensional frequencies $\eta(= a\omega/c_S)$ of spheroidal oscillations of a uniform elastic shell with liquid core, material constants being given by (4.2.2) and (5.3.2).

where suffixes o and i indicate mantle and core respectively. Where the core is liquid, μ_i must be put equal to zero and the second equation of (5.3.1) discarded.

Putting nine analytical solutions into the nine boundary conditions, we obtain the frequency equation as a consistency condition. This equation takes the form of the vanishing of the determinant of of the ninth order. This could be solved to obtain numerical values of the non-dimensional frequency $\eta = \omega a/c_{So}$. However, manipulation of the ninth-order determinant is troublesome and short-sighted. A better method of solving the differential equations (5.2.4) under the appropriate boundary conditions is recommended for numerical computations in Chapter 6.

As an Earth-model, we choose that given by (4.2.2). However, in the present problem we need not only the ratio of densities ρ_i/ρ_o, but the absolute value of the densities. These values are assumed to be

$$\rho_i = 10.171 \text{ g/cm}^3, \qquad \rho_o = 4.6232 \text{ g/cm}^3, \tag{5.3.2}$$

which have been chosen so as to give the mean density of the model sphere as 5.52 g/cm^3, which is the mean density of the Earth.

The non-dimensional frequencies calculated for this model are shown in Fig. 5.2 as functions of Legendre degree l, for both gravitating and non-gravitating spheres. Here again the figure shows that the effect of gravity is not simple. For purely radial oscillations ($l = 0$), gravity lengthens the period. For the fundamental mode $n = 0$ for any $l > 0$, the effect of gravity is to shorten the eigenperiod, a natural conclusion from simple physical considerations; tightening the Earth by self-gravitation decreases the natural period. For other modes, gravity involves complicated effects. It is seen from Fig. 5.2 that the effect of gravity is large for the special mode $n = n_s$, which has large amplitudes near the core–mantle boundary, a characteristic associated with Stoneley waves along a boundary between the different media. For these modes gravity shortens the eigenperiod. For all other modes the effect of gravity decreases as the Legendre degree l increases. In other words, gravity affects short waves less than long.

6

CALCULATION OF EIGENFREQUENCIES AND EIGENFUNCTIONS FOR REALISTIC EARTH-MODELS

6.1 Methods and processes of calculation of eigenfrequencies and eigenfunctions

The numerical calculation of eigenfrequencies and eigenfunctions is not easy, even for the simplest example – the toroidal oscillation of a uniform elastic sphere. The solution of the frequency equation (2.3.11) requires detailed knowledge of spherical Bessel functions and the numerical process is troublesome. Consequently the study of free oscillations of an elastic sphere has depended on the development of high-speed electronic computers. Special methods, suitable for computations by electronic computers, are described below.

6.1.1 *Rayleigh's Principle*

Let us consider a finite system performing small harmonic free vibrations in a certain normal mode with angular frequency ω. The kinetic energy (T) and potential energy (V) of the whole system, averaged over one period, can be shown to be equal. The former has a factor ω^2, so that ω^2 may be expressed as

$$\omega^2 = V/T', \tag{6.1.1}$$

where T' is T/ω^2. One form of Rayleigh's Principle (Rayleigh, 1896, § 88, Jeffreys & Jeffreys, 1956, § 4.084) states that the right-hand side of (6.1.1) is stationary for small variations of the displacements from the exact solution. In other words, calculation of the right-hand side of (6.1.1), using an approximate trial solution with first-order deviation from the exact eigenfunction, gives the eigenfrequency accurate to the first order. It is clear that the success of this method depends on making a good guess for the trial solution. When this method is applied to the gravest value of ω, the estimate approaches the exact value from above. Jobert, in some of the earliest attempts (1956, 1957a, b), employed Rayleigh's Principle to give eigenfrequencies for a simple elastic sphere and for a realistic earth model (see Chapter 1). The method estimates the eigenfrequency, but does not provide the eigenfunction. It has therefore been superseded by the Rayleigh–Ritz method, which we describe in § 6.1.2 below.

6.1.2 *Variational method*

A more powerful variational method is formulated as follows (Jeffreys & Jeffreys, 1956, § 6.08, Alterman, Eyal & Merzer, 1974, Sokolnikoff, 1956, §§ 112, 113). We work with a functional

$$I = \int f\{\mathbf{u}(\mathbf{r}), \mathbf{r}, \omega\}\,\mathrm{d}V, \tag{6.1.2}$$

where $\mathbf{r}$ is the position vector, ω an eigenfrequency, and $\mathbf{u}(\mathbf{r})$ the corresponding eigenfunction; the integral is taken over the whole system. The function f is constructed in such a way that the Euler–Lagrange equations corresponding to stationary I are the equations of motion of the system. Thus, selection of $\mathbf{u}$ to make I stationary is equivalent to solution of the equations of motion.

In the Rayleigh–Ritz variational method, linearly independent trial functions $\mathbf{u}_i(\mathbf{r})$ are chosen so that each satisfies the boundary conditions; then a trial solution $\mathbf{U}_k(\mathbf{r})$ is constructed as a linear combination of $\mathbf{u}_i$, that is,

$$\mathbf{U}_k = \sum_{i=1}^{k} a_i \mathbf{u}_i, \tag{6.1.3}$$

where a_i are parameters which are to be determined so as to make the integral I stationary. On inserting (6.1.3) in (6.1.2) we obtain $I(a_1, a_2, \ldots, a_k, \omega)$ which can be made stationary by calculating the values of a_i from the system of equations

$$\frac{\partial I}{\partial a_i} = 0, \qquad i = 1, 2, \ldots, k. \tag{6.1.4}$$

In our problem, f is a quadratic function of $\mathbf{u}$, and a linear function of ω^2, and I is consequently a homogeneous quadratic function of the a_i, whose coefficients are linear functions of ω^2. Thus (6.1.4) is a set of homogeneous linear equations in a_i, and requires, for consistency, the vanishing of the determinant of coefficients. This gives an equation of degree k for ω^2, the roots of which may be shown to give estimates of the first k eigenfrequencies of vibration. To each ω corresponds a set of coefficients $a_i(\omega)$. The corresponding $\mathbf{u}_i$ will be estimates of the eigenfunctions, but each will be contaminated by contributions from other modes. To obtain better estimates of the eigenfunctions we must aim to construct a set of orthogonal trial functions, each satisfying the boundary conditions. It seems plausible that $\mathbf{U}_k$ will be a satisfactory approximation to an exact solution when the choice of trial functions is appropriate and the number of parameters a_i is made sufficiently large; this can be shown to be true. Thus, if the trial functions are skilfully chosen, we can obtain reasonably accurate numerical solutions using less computing time than is taken by direct numerical solution of the differential equations of motion.

Takeuchi (1959) applied this method to calculation of the first three eigenfrequencies of toroidal oscillation of a realistic earth-model. He adopted Bullen's model (Bullard, 1954). Convergence was very rapid and $k = 3$ gave satisfactory results. For the mode $l = 2$ and $n = 0$, he obtained the eigenperiod of 43.4 min, which agreed well with 43.54 min as given by Jobert (1959).

In another early paper, Pekeris & Jarosch (1958) applied a variational method to the spheroidal oscillations of a realistic self-gravitating Earth-model (Bullen B). This example, however, used only $k = 1$. They assumed a trial eigenfunction $\mathbf{u}_1$ with vertical and transverse components given by polynomial expressions in r, r^3 and r^5, where r is the radial distance; the coefficients in the polynomials were chosen so as to satisfy the boundary conditions. The trial function for the gravitational potential Y was obtained by integrating the last equation of (5.2.4), using the trial eigenfunctions and assumed values of ρ_0. Thus they obtained 53.0 min for the period of the mode $l = 2$ and $n = 0$; this is near to values now generally obtained.

The variational method has been formalised in other ways, and we usually have to solve a set of simultaneous equations like (6.1.4) to get the eigenfrequencies of higher overtones. Eigenfrequencies, even those of higher modes, which depend on assuming a good set of orthogonal trial functions, have been formed successfully with adequate accuracy by variational methods.

6.1.3 *Thomson–Haskell matrix method*

In the Thomson–Haskell method, a model is constructed of uniform layers, that is, uniform spherical shells; the thinner each shell, the better the approximation. The model may be called partially realistic. This method, first formulated by Thomson (1950) and extended by Haskell (1953) and Matumoto (1953), can be used provided an analytical solution for a single uniform elastic shell is known. Gilbert & MacDonald (1960) applied the method to the toroidal oscillation of a realistic Earth-model as follows.

Time-harmonic solutions of the equation of motion in each spherical shell are given by (2.2.29), which we reproduce in (6.1.5):

$$\left.\begin{aligned} u &= 0, \\ v &= U_{ml}\frac{m}{\sin\vartheta}P_l^m(\cos\vartheta)\,{}^{\sin}_{(-\cos)}\,m\varphi\mathrm{e}^{\mathrm{i}\omega t}, \\ w &= U_{ml}\partial_\vartheta P_l^m(\cos\vartheta)\,{}^{\cos}_{\sin}\,m\varphi\mathrm{e}^{\mathrm{i}\omega t}, \end{aligned}\right\} \tag{6.1.5}$$

where

$$U_{ml} = \{B_{ml}j_l(kr) + B'_{ml}y_l(kr)\},$$

and B_{ml} and B'_{ml} are constant multipliers to be determined from the boundary conditions. The stress components on the spherical surface are

$$\left.\begin{aligned} \widehat{rr} &= 0, \\ \widehat{r\vartheta} &= \mu T_{ml}\frac{m}{\sin\vartheta}P_l^m(\cos\vartheta)\,{}^{\sin}_{(-\cos)}\,m\varphi\,\mathrm{e}^{\mathrm{i}\omega t}, \\ \widehat{r\varphi} &= \mu T_{ml}\partial_\vartheta P_l^m(\cos\vartheta)\,{}^{\cos}_{\sin}\,m\varphi\,\mathrm{e}^{\mathrm{i}\omega t}, \end{aligned}\right\} \tag{6.1.6}$$

where

$$T_{ml} = B_{ml}\left\{\dot{j}_l(kr) - \frac{1}{r}j_l(kr)\right\} + B'_{ml}\left\{\dot{y}_l(kr) - \frac{1}{r}y_l(kr)\right\},$$

and dot represents $\mathrm{d}/\mathrm{d}r$.

Since the boundary conditions on each surface are the continuity of displacement and stress, and since the factors in ϑ and φ are continuous while the factor in r has identical forms for v and w and for $\widehat{r\vartheta}$ and $\widehat{r\varphi}$, it is sufficient to consider the factors in r of w and

$\widehat{r\varphi}$. Then

$$\begin{bmatrix} U_{ml} \\ \\ T_{ml} \end{bmatrix} = \begin{bmatrix} j_l(kr) & y_l(kr) \\ \\ \dot{j}_l(kr) - \dfrac{1}{r} j_l(kr) & \dot{y}_l(kr) - \dfrac{1}{r} y_l(kr) \end{bmatrix} \begin{bmatrix} B_{ml} \\ \\ B'_{ml} \end{bmatrix}. \tag{6.1.7}$$

Now consider a model consisting of N shells over a liquid core. Let the subscript i refer to the ith spherical shell, and let r_i and r_{i+1} be the radii of the inner and outer surfaces of the ith shell. The radius of the earth is $r_{N+1} = a$, and the radius of the core is $r_1 = b$. The core is assumed to be perfect-liquid. Equations (6.1.7), applied to the ith shell, are abbreviated to

$$\begin{bmatrix} U_{ml,i}(r) \\ T_{ml,i}(r) \end{bmatrix} = \mathscr{A}_i(r) \begin{bmatrix} B_{ml,i} \\ B'_{ml,i} \end{bmatrix}, \tag{6.1.8}$$

where $\mathscr{A}(r)$ is the 2×2 matrix in (6.1.7) and $\mathscr{A}_i(r)$ is here a function of $k_i = \omega / c_{Si}$ and hence of the density ρ_i and the rigidity μ_i of the ith shell. The inverse of (6.1.8) is

$$\begin{bmatrix} B_{ml,i} \\ B'_{ml,i} \end{bmatrix} = \mathscr{A}_i^{-1}(r) \begin{bmatrix} U_{ml,i}(r) \\ T_{ml,i}(r) \end{bmatrix}, \tag{6.1.9}$$

provided $\mathscr{A}_i$ is not singular. Combining (6.1.8) for $r = r_i$ and (6.1.9) for $r = r_{i+1}$, $B_{ml,i}$, $B'_{ml,i}$ being the same at the two boundaries, we have

$$\begin{bmatrix} U_{ml,i}(r_i) \\ T_{ml,i}(r_i) \end{bmatrix} = \mathscr{A}_i(r_i) \mathscr{A}_i^{-1}(r_{i+1}) \begin{bmatrix} U_{ml,i}(r_{i+1}) \\ T_{ml,i}(r_{i+1}) \end{bmatrix}, \tag{6.1.10}$$

which connects the stresses and displacements at levels $r = r_i$ and $r = r_{i+1}$. We now introduce 2×2 matrices $\mathscr{B}_i$ and $\mathscr{C}$, defined by

$$\left.\begin{aligned} \mathscr{A}_i(r_i)\mathscr{A}_i^{-1}(r_{i+1}) = \mathscr{B}_i, \\ \mathscr{B}_1 \mathscr{B}_2 \ldots \mathscr{B}_N = \mathscr{C}. \end{aligned}\right\} \tag{6.1.11}$$

Since the boundary conditions are

$$\begin{bmatrix} U_{ml,i}(r_{i+1}) \\ T_{ml,i}(r_{i+1}) \end{bmatrix} = \begin{bmatrix} U_{ml,i+1}(r_{i+1}) \\ T_{ml,i+1}(r_{i+1}) \end{bmatrix}, \tag{6.1.12}$$

at each boundary, we have

$$\begin{bmatrix} U_{ml,1}(r_1) \\ T_{ml,1}(r_1) \end{bmatrix} = \mathscr{C} \begin{bmatrix} U_{ml,N}(r_{N+1}) \\ T_{ml,N}(r_{N+1}) \end{bmatrix}. \tag{6.1.13}$$

Since the stress vanishes on the core–mantle interface and on the outer surface of the Earth, we have

$$\begin{bmatrix} U_{ml,1}(b) \\ 0 \end{bmatrix} = \mathscr{C} \begin{bmatrix} U_{ml,N}(a) \\ 0 \end{bmatrix}, \tag{6.1.14}$$

or, if we write

$$\mathscr{C} \equiv \begin{bmatrix} C_{11} & C_{12} \\ C_{21} & C_{22} \end{bmatrix}, \tag{6.1.15}$$

and

$$\left.\begin{aligned} U_{ml,1}(b) &= C_{11}\, U_{ml,N}(a), \\ 0 &= C_{21}\, U_{ml,N}(a). \end{aligned}\right\} \tag{6.1.16}$$

Therefore

$$C_{21} = 0; \tag{6.1.17}$$

this equation gives the eigenfrequencies of toroidal oscillations. If we define k as $k_N = \omega/c_{SN}$, C_{21} is a function of k, and we can search for the roots of (6.1.17) numerically as functions of k. Gilbert & MacDonald (1960) represented the Earth by 30 shells and calculated eigenfrequencies of modes ($l = 1$–18, $n = 0$–3) for both Gutenberg and Jeffreys–Bullen Earth-models. The eigenfunction corresponding to a designated eigenvalue was obtained as a set of analytical solutions (6.1.5) for all layers, joined smoothly at the interfaces.

The Thomson–Haskell method introduced here is exact for an elastic sphere consisting of uniform shells, but the solution applies to the real Earth only as far as the model of uniform layers fits the real Earth. The frequency equation for spheroidal oscillations can be formulated in the same way, using 4×4 matrices in place of the 2×2 matrices above. The introduction of gravity raises the order of the matrices to 6×6 and complicates the elements of the matrices. For these reasons the speed of computation by the Thomson–Haskell technique is not much greater, even for the simplest examples, than that of the direct method which will be explained in the next subsection. Dividing the Earth-model into more spherical shells provides a better approximation to the structure of the real Earth; but on the other hand it also involves more matrix manipulations and consequent loss of numerical accuracy.

6.1.4 *Numerical integration*

A straightforward method of solution of the eigenvalue problem is by numerical integration of the equations of motion, taking care to satisfy the boundary conditions. Before we deal with the method of numerical integration, let us fill a gap in our analysis by introducing the equations of motion of toroidal oscillation of a radially *heterogeneous* elastic sphere.

Assuming the material parameters ρ, λ and μ to be functions of r alone, we may reasonably guess, from analogy with the solution (2.2.29) for a uniform elastic sphere, that the solution has the form

$$\left.\begin{aligned} u &= 0, \\ v &= B_{ml} W(r) \frac{m}{\sin\vartheta} P_{l}^{m} {}_{(-\cos)}^{\ \sin} m\varphi\, e^{i\omega t}, \\ w &= B_{ml} W(r) \partial_\vartheta P_{l}^{m} {}_{\sin}^{\cos} m\varphi\, e^{i\omega t}, \end{aligned}\right\} \tag{6.1.18}$$

where $W(r)$ is a function of r which is to be determined. Remembering that gravity has no effect on toroidal oscillations and that the dilatation Δ is to be zero, and putting (6.1.18) into (5.1.12), we obtain the field equation for $W_l(r)$:

$$\mu \ddot{W}_l + (\dot{\mu} + 2\mu/r)\dot{W}_l - \{\dot{\mu}/r + l(l+1)\mu/r^2 - \omega^2 \rho_0\} W_l = 0, \tag{6.1.19}$$

where the dot represents d/dr, as before. The boundary condition at the surface $r = a$ is

$$\dot{W}_l(a) - W_l(a)/a = 0,$$

and the conditions at any spherical interface are

$$\left.\begin{aligned} [\mu\{\dot{W}(r) - W(r)/r\}]_o &= [\mu\{\dot{W}(r) - W(r)/r\}]_i, \\ [W(r)]_o &= [W(r)]_i, \end{aligned}\right\} \tag{6.1.20}$$

where the suffixes o and i refer to the upper and lower sides of the interface.

These equations and boundary conditions correspond to (5.2.4), (5.2.7), (5.2.8) and (5.3.1) for spheroidal oscillations.

Toroidal oscillations Thus for toroidal oscillations we have one second-order linear differential equation to be solved for the unknown $W(r)$. Numerical integration may be carried out starting either from the surface or from the core–mantle boundary. In either case the solution has to satisfy the boundary condition (6.1.20) at each surface.† Thus the marching solution must be carried out starting with one adjustable parameter, which will have to be determined to make the final value satisfy the correct end-condition. This adjustable parameter may be the frequency: then its value determined to satisfy the end-condition will be an eigenfrequency. It is advantageous to carry out numerical integration starting with the Legendre degree $l = 1$ and proceeding to successively larger values of l; and to begin, for each value of l, with ${}_0\omega_{l-1}$, the gravest eigenfrequency for the degree $l - 1$, to which is added suitably chosen $\Delta\omega$ at each step in the finding of ${}_0\omega_l$. Having obtained eigenfrequencies for fundamental modes ($n = 0$) we can proceed to those of the overtones ($n \geq 1$).

If we try to integrate (6.1.19) numerically step by step, we run up against a major source of inaccuracy. At any step, in order to obtain the value of $W(r)$ for the next step, we must calculate $\dot{W}$ and $\ddot{W}$ numerically. But numerical differentiation loses accuracy. So in order to avoid the second differentiation we replace (6.1.19) by two simultaneous first-order differential equations as follows (Alterman *et al.*, 1959):

Put

$$\left.\begin{aligned} y_1 &= W(r), \\ y_2(=\widehat{r\varphi}) &= \mu\{\dot{W}(r) - W(r)/r\}. \end{aligned}\right\} \tag{6.1.21}$$

Then we have

$$\left.\begin{aligned} \dot{y}_1 &= y_1/r + y_2/\mu, \\ \dot{y}_2 &= \{(l-1)(l+2)\mu/r^2 - \rho_0\omega^2\}y_1 - 3y_2/r. \end{aligned}\right\} \tag{6.1.22}$$

Note that we have thus avoided also the computation of derivatives

†If we consider not a shell but a solid sphere, the condition at the inner boundary must be replaced by a condition of regularity at the centre.

of μ. The boundary condition at the surface is

$$y_2 = 0, \tag{6.1.23}$$

and the conditions at interface between two different elastic media are replaced by the continuity of y_1 and y_2, that is

$$\left.\begin{aligned}[y_1]_o &= [y_1]_i,\\ [y_2]_o &= [y_2]_i.\end{aligned}\right\} \tag{6.1.24}$$

If the core is perfect-liquid, we obtain the eigenfrequency ω and eigenfunction y_1 by solving the simultaneous first-order differential equations (6.1.22) with two unknowns, y_1 and y_2, under the conditions

$$y_2 = 0 \qquad \text{at } r = a, r = b. \tag{6.1.25}$$

Spheroidal oscillations For spheroidal oscillations we have three second-order linear differential equations (5.2.4) to be solved for the unknowns $U(r)$, $V(r)$ and $Y(r)$. Here the marching solution, starting from, say, the surface, must satisfy boundary conditions at the surface and must contain three adjustable parameters which are to be determined so as to satisfy boundary conditions at the further end-point (in this case, the centre). These parameters may be the relative amplitudes $V(a)/U(a)$ and $Y(a)/U(a)$ at the surface, together with a frequency ω. Thus the numerical solution involves a search for a point in the three-dimensional space $\{\omega, V(a)/U(a), Y(a)/U(a)\}$.†

The differential equations (5.2.4) for spheroidal oscillations can be reorganised to reduce numerical differentation by the transformation

$$\left.\begin{aligned}
y_1 &= U,\\
y_2(&= \widehat{rr}) = \lambda\Delta + 2\mu\dot{U},\\
y_3 &= V,\\
y_4(&= \widehat{r\vartheta}) = \mu\{\dot{V} - (V - U)/r\},\\
y_5 &= Y,\\
y_6 &= \dot{Y} - 4\pi G\rho_0 U.
\end{aligned}\right\} \tag{6.1.26}$$

†The three corresponding parameters when we start numerical integration from the centre will be described below.

We obtain

$$
\left.\begin{aligned}
\dot{y}_1 &= -\frac{2\lambda}{\lambda+2\mu}\frac{y_1}{r} + \frac{1}{\lambda+2\mu}y_2 + \frac{\lambda l(l+1)}{\lambda+2\mu}\frac{y_3}{r}, \\
\dot{y}_2 &= \left\{-\rho_0\omega^2 r^2 - 4g_0\rho_0 r + \frac{4\mu(3\lambda+2\mu)}{\lambda+2\mu}\right\}\frac{y_1}{r^2} - \frac{4\mu}{\lambda+2\mu}\frac{y_2}{r} \\
&\quad + \left\{l(l+1)g_0\rho_0 r - \frac{2l(l+1)\mu(3\lambda+2\mu)}{\lambda+2\mu}\right\}\frac{y_3}{r^2} \\
&\quad + l(l+1)\frac{y_4}{r} - \rho_0 y_6, \\
\dot{y}_3 &= -\frac{y_1}{r} + \frac{y_3}{r} + \frac{y_4}{\mu}, \\
\dot{y}_4 &= \left\{g_0\rho_0 r - \frac{2\mu(3\lambda+2\mu)}{\lambda+2\mu}\right\}\frac{y_1}{r^2} - \frac{\lambda}{\lambda+2\mu}\frac{y_2}{r} \\
&\quad + \left[-\rho_0\omega^2 r^2 + \frac{2\mu}{\lambda+2\mu}\{\lambda(2l^2+2l-1) \right. \\
&\quad \left. + 2\mu(l^2+l-1)\}\right]\frac{y_3}{r^2} - 3\frac{y_4}{r} - \rho_0\frac{y_5}{r}, \\
\dot{y}_5 &= 4\pi G\rho_0 y_1 + y_6, \\
\dot{y}_6 &= -4\pi G\rho_0 l(l+1)\frac{y_3}{r} + l(l+1)\frac{y_5}{r^2} - 2\frac{y_6}{r}.
\end{aligned}\right\} \quad (6.1.27)
$$

The boundary conditions at the surface are now

$$
\left.\begin{aligned}
y_2 &= 0, \\
y_4 &= 0, \\
y_6 &= -(l+1)y_5/a
\end{aligned}\right\} \text{at } r = a. \quad (6.1.28)
$$

By analogy with a uniform elastic sphere (cf. (5.2.11)), for the regularity conditions at the centre to be satisfied it is sufficient that

$$
y_1 = 0, \qquad y_3 = 0, \qquad y_5 = 0 \qquad \text{at } r = 0, \quad (6.1.29)
$$

except when $l = 1$. When $l = 1$, (6.1.29) must be replaced by

$$
y_1 = y_3, \qquad \dot{y}_1 = 0, \qquad y_5 = 0. \quad (6.1.30)
$$

At an interface between different media, $y_1, \ldots, y_6$ must be continuous. Solving (6.1.27) for the six unknowns $y_1, \ldots, y_6$ under the boundary conditions (6.1.28–30) gives the eigenfrequency and the eigenfunctions $U(r)$, $V(r)$ and $Y(r)$ to within a constant multiplier.

If we start integration from the outer surface, it is convenient to put $y_1(a) = 1$. We then assume values of ω, $V(a)$, $Y(a)$ and determine the first derivatives of U, V, Y so as to satisfy the boundary conditions at $r = a$. Then using the differential equations and Taylor's expansion we obtain corresponding values at $r = a - \delta r$. The process is repeated to give values at $r = a - 2\delta r,\ a - 3\delta r, \ldots$. The process is found to diverge at the centre, or at some intermediate radius, unless ω is an eigenvalue. From the characteristics of the eigenfunctions of simple spheres, it is to be expected that the eigenfunctions here become negligibly small below certain depths. Therefore, instead of the conditions at $r = 0$, it is more convenient to impose conditions that the unknown functions U, V, Y vanish at a certain depth r_c at which it is estimated that the eigenfunctions fall below an assigned small amplitude. From experience gained from lower modes, we can confidently fix a suitable depth r_c for each higher mode. Changing the surface values of V and Y and eigenfrequency ω, we search for a particular combination of (V, Y, ω) satisfying the assumed conditions at $r = r_c$. This gives the eigenfrequency and eigenfunctions. For modes with larger l, in which the non-zero displacement is compressed into the shallower part of the sphere, this method of starting numerical integration from the surface is very efficient. Numerical integration shows that the depth below which the displacement is assumed to be zero (a main parameter affecting the accuracy of the eigenfrequency) affects, at most, the fourth digit of the eigenfrequency.

If we start integration from the centre of the sphere, we assume that the first derivative of each eigenfunction vanishes at the centre. (This can be justified by considering a small uniform sphere of radius δr, since we assume that material parameters are continuous at the centre.) But neither this character nor the regularity requirement at the centre gives means to calculate values at $r = \delta r$ as far as numerical solution of the differential equations goes. In order to obtain values at $r = \delta r$, we have to assume that the conditions near the centre are adequately represented by a uniform small sphere of radius δr. Then we can calculate the values at $r = \delta r$ using analytic solutions within this uniform sphere. After that, we can proceed step by step to the surface. Generally, the three boundary conditions at the surface will not be satisfied by our marching solution. For a uniform liquid small sphere of radius δr, the three independent solutions with constant multipliers R, S and T in (5.2.11) reduce to two, that is, one with T and one other. We adopt (i) the ratio of these two independent solutions; (ii) the eigenfrequency ω; and (iii) V of the mantle at $r = b$ (see later) as the three adjustable unknown parameters.

Finally we search for a suitable combination of the adjustable parameters to satisfy the three boundary conditions at the surface. For details, particularly for the choice of two independent solutions in a small sphere with radius δr, refer to Takeuchi & Saito (1972).

At the core–mantle boundary, the condition of continuity of tangential displacement disappears when the core is liquid, and that of continuity of tangential stress is not adequate for the continuation of numerical integration across the core–mantle interface. In the core, μ and $d\mu/dr$ are zero. Thus the equations of motion (5.2.4) reduce to two equations with unknowns U and Y. If we have started numerical integration from the surface, starting values at the core boundary needed in order to continue numerical integrations are those of $U, \dot{U}, Y$ and $\dot{Y}$ at $r = b$. Among these, U, Y and $\dot{Y}$ are obtained from the first, fifth and sixth relations of (5.3.1). The starting value of $\dot{U}$ can be obtained by solving the third equation of (5.3.1) with the second equation (5.2.4) as two simultaneous equations for unknowns $\dot{U}$ and V (of the core) at $r = b$.

If we start numerical integration from the centre, we need starting values of U, V, Y, $\dot{U}$, $\dot{V}$, $\dot{Y}$ at the base of the mantle. Of these, U, Y, $\dot{Y}$ are available from core values and boundary conditions at $r = b$. We may then give to V an arbitrary value at $r = b$ and obtain $\dot{U}$ and $\dot{V}$ at the base of the mantle from the third and fourth equations of (5.3.1). Thus the value $V(b)$ in the mantle is our third adjustable parameter.

For solving the first-order simultaneous equations (6.1.22) or (6.1.27), the Runge–Kutta method of numerical integration is recommended (see e.g. Scarborough, 1962, § 116, Hornbeck, 1975 § 9.4, Hall & Watt, 1976, Chapter 5). For equations (6.1.19) and (5.2.4), both Milne's method (Milne, 1949, Chapter 5) and the Runge–Kutta method are effective.

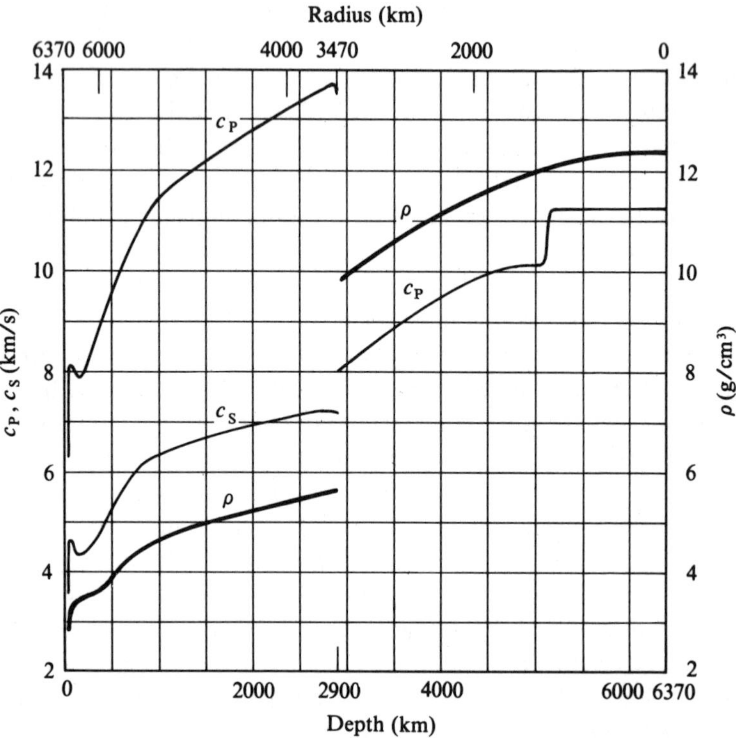

Fig. 6.1. P- and S-wave velocities as functions of depth (Gutenberg), and density as function of depth (Bullen A′).

6.2 Eigenfrequencies and eigenfunctions for a realistic Earth-model

We now show the results of the computations described above, applied to an Earth-model with Gutenberg velocities and Bullen A′ density distribution. These are shown in Fig. 6.1 as functions of depth. A crust was characterised by a thickness of 32 km, $c_S = 3.55$ km/s, $c_P = 6.30$ km/s and density 2.84 g/cm^3 (Usami, Satô & Landisman, 1966, Satô, Usami & Landisman, 1968).

There exist more recent Earth-models, which have been designed to give the best fit to data now available. But since these differ only in detail from our chosen model, for which computations and diagrams were readily accessible to us, and since we are concerned rather with basic results than with small modifications, we use Gutenberg–Bullen A′.

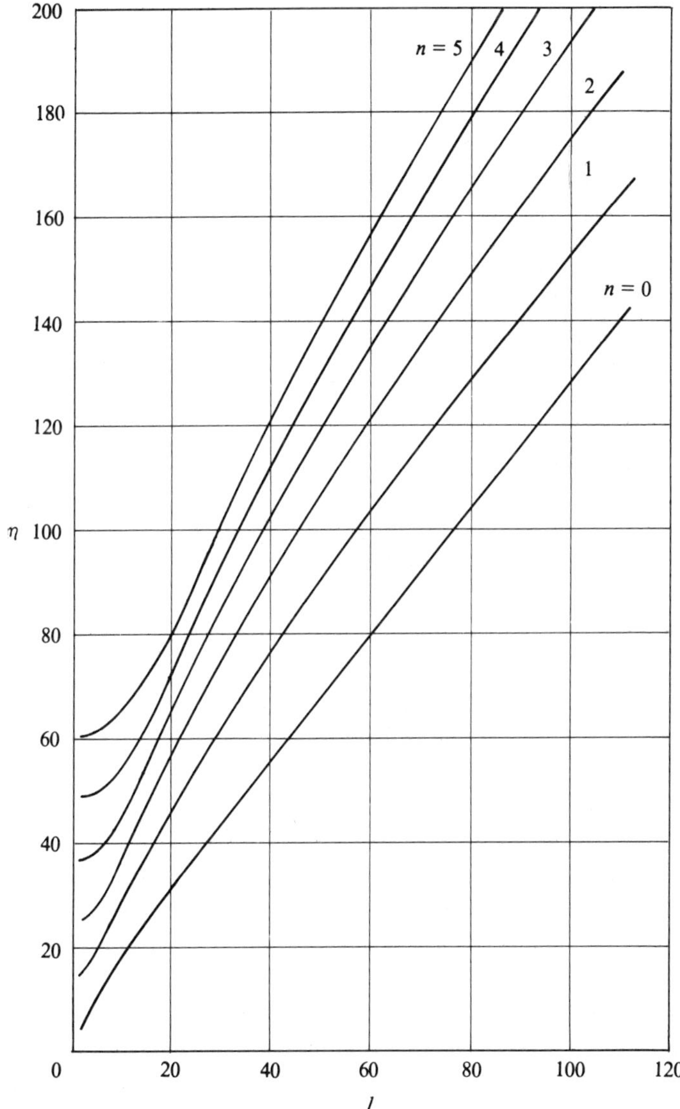

Fig. 6.2. Non-dimensional frequency η ($= \omega a/c_{\mathrm{S}_0}$) of toroidal oscillations for Gutenberg–Bullen A′ Earth-model. (Note that the continuous curves join up points corresponding to modal frequencies. Only those points, and not the intervening curves, represent physical reality.)

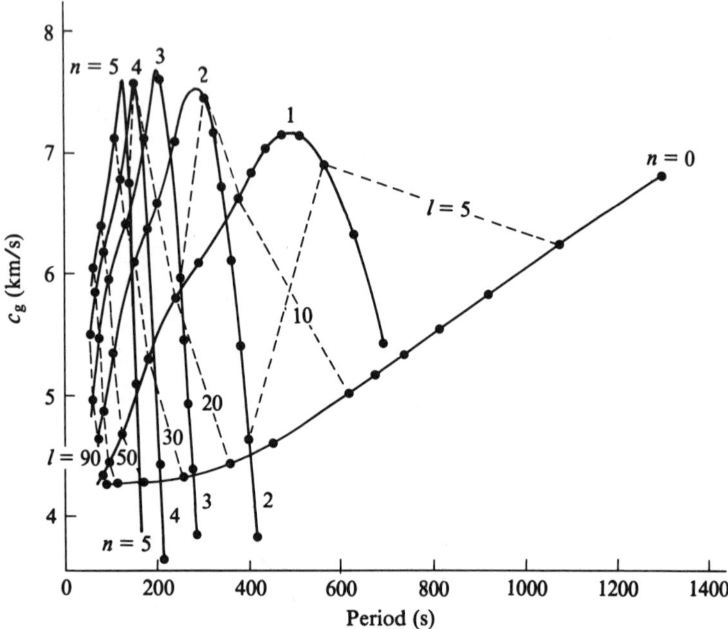

Fig. 6.3. Group-velocity, as function of period, of toroidal oscillations for Gutenberg–Bullen A′ Earth-model. Only the points marked correspond to eigenfrequencies. Solid curves link eigenperiods for the same overtone and different degree l. Broken dashed lines join eigenperiods for the same degree l and different overtones.

6.2.1 *Toroidal oscillations*

The non-dimensional frequencies $\eta \doteq \omega a/c_{\mathrm{So}}$ (c_{So} is S-wave velocity in the crust) are shown in Fig. 6.2 as functions of Legendre degree l and overtone number n. The general pattern of these curves is quite similar to that for a uniform shell (Fig. 3.1).

The group-velocity calculated from (2.5.9) as a function of period is shown in Fig. 6.3. The curve joining eigenfrequencies of fundamental modes is monotonic. Those for higher overtones exhibit narrow maxima whose values lie in the vicinity of 7.5 km/s for all $n \geq 1$. The calculated group-velocity maxima, which occur at successively shorter periods as n increases, have been shown to be related to a wave which grazes the core (cf. § 3.2 and Satô, Landisman & Ewing, 1960).

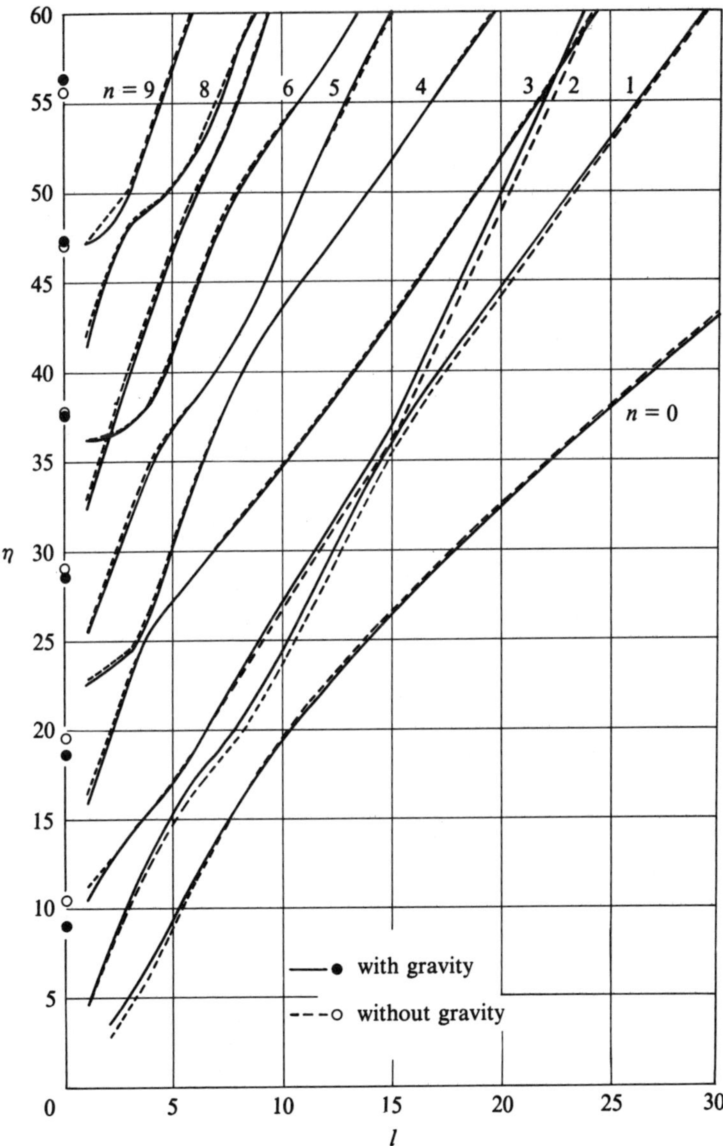

Fig. 6.4. Non-dimensional frequency of spheroidal oscillations for Gutenberg–Bullen A′ Earth-model. (See note on Fig. 6.2.)

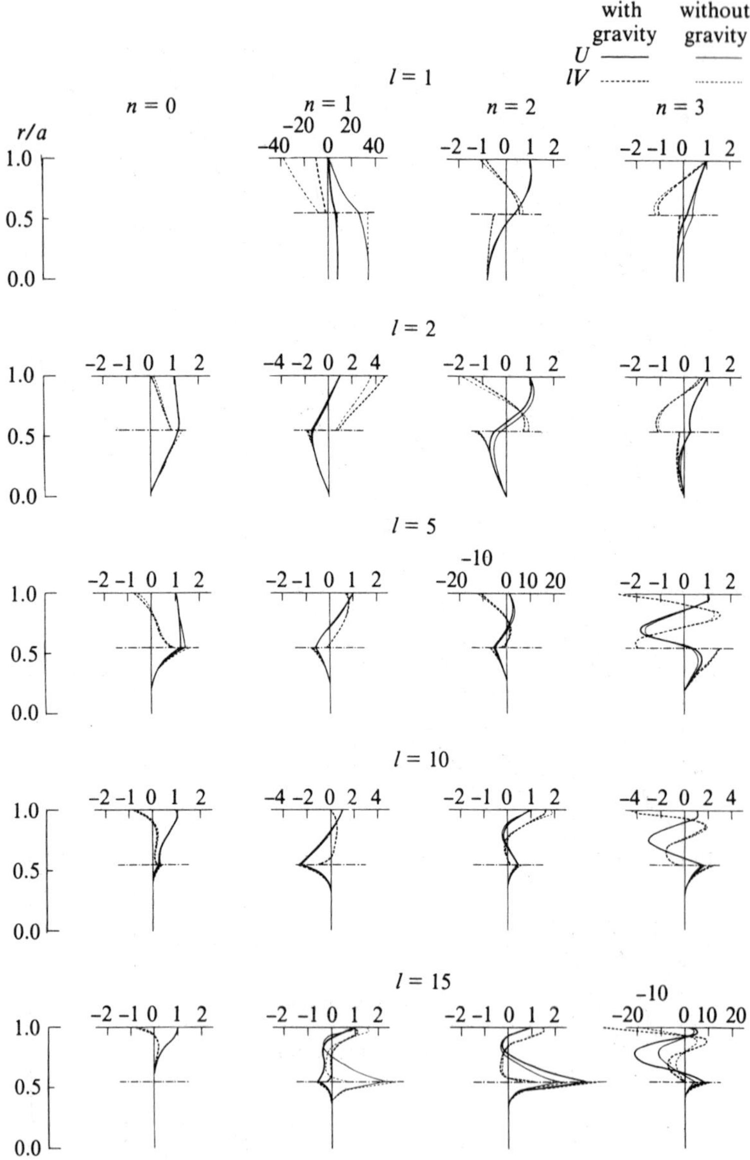

Fig. 6.5. Radial distributions of radial and transverse components of displacement. Thick curves refer to gravitating sphere and thin curves to non-gravitating sphere. Solid lines give radial and dotted lines transverse components. Radial displacement at the free surface is taken to be unity. Chain lines indicate core–mantle boundary.

6.2.2 *Spheroidal oscillations*

The non-dimensional frequency $\eta(=\omega a/c_{S_0})$ as function of l and n is shown in Fig. 6.4 for self-gravitating and non-gravitating Earth-models. (Note that the range of l and η is smaller than that in Fig. 6.2 for toroidal oscillations.) Transition segments from one overtone to the next higher are seen in the figure (for instance, at $n = 2$, $l = 15$–22). Near the beginning and end-points of transition segments, we can see quasi-junctions between curves having adjacent Legendre degrees. Modes for $l = 15$ on the transition segment of $n = 2$ near the curve for $n = 1$ are found to display large amplitudes near the core–mantle interface (Fig. 6.5), suggesting that these modes are closely related to the Stoneley wave along the core–mantle interface.

Minor transition segments are seen in Fig. 6.4 for smaller values of l. Since these appear also for a uniform sphere (Odaka & Usami, 1978), they cannot be due to boundary waves along interfaces existing in real Earth-models, but are associated with waves reflected at the surface.

In Fig. 6.4, solid lines (and solid circles on $l = 0$) refer to the self-gravitating sphere and broken lines (and open circles on $l = 0$) to the non-gravitating sphere. The effect of gravity on the non-dimensional frequency depends on the mode. For fundamental modes with $l > 9$, gravity lengthens the eigenperiod. The effect of gravity is largest for transition segments ($n = 1, l = 7$–14 and $n = 2, l = 13$–22), that is, for modes related to Stoneley waves along the core–mantle interface. The effect of gravity on overtones with $l = 0$ is not as simple as for a uniform sphere or for a sphere with uniform mantle and core. From this figure it appears that for larger l gravity lengthens the eigen-period for lower overtones and shortens for higher.

Fig. 6.5 shows eigenfunctions of some modes for both gravitating and non-gravitating spheres. We omit modes with $l = 1$ and $n = 0$, since they are not excited in free oscillations (cf. Chapter 2).

6.3 Comparison of eigenfrequencies for different Earth-models

For the calculation of eigenfrequencies for the toroidal and spheroidal oscillations of a realistic Earth-model, the distributions of (ρ, λ, μ), or alternatively (ρ, c_P, c_S) as functions of radius must be given. In earlier work the standard velocity distributions used were

Table 6.1. *Periods of observed and theoretical toroidal modes in minutes. Fundamental mode* ($n = 0$).

	Observed period		Theoretical period			
Mode l	1964 Alaskan Earthquake	Mean of Chilean data (Press, 1964)	Jeffreys–Bullen A (Satô *et al.*, 1960)	Jeffreys–Bullen B (Pekeris *et al.*, 1961b)	Gutenberg–Bullen A (Pekeris *et al.*, 1961b)	UTD 124A′ (Dziewonski & Gilbert, 1972)
6	15.34	15.36	15.28	15.54	15.36	15.41
7	13.73	13.53	13.37	13.72	13.57	13.63
8	12.27	12.25	12.13	12.33	12.22	12.26
9	10.96	11.14	11.05	11.23	11.15	11.19
10	10.58	10.32	10.20	10.34	10.28	10.31
11	9.689	9.614	9.50	9.59	9.55	9.58
12	8.937	8.975	8.86	8.95	8.93	8.95
13	8.368	8.379	8.33	8.40	8.39	8.41
14	7.876	7.942	7.86	7.92	7.93	7.94
15	–	7.527	7.43	7.50	7.51	7.52
16	7.081	7.161	7.07	7.12	7.15	7.15
17	6.786	6.819	6.73	6.78	6.81	6.82
18	6.528	6.502	6.45	6.48	6.52	6.52
19	6.262	6.246	6.18	6.20	6.24	6.24
20	6.016	6.006	5.93	5.95	5.99	5.99
21	5.796	5.769	5.70	5.72	5.77	5.76

those of Gutenberg and Jeffreys, and standard density distributions those of Bullen and Bullard. Since Bullen and Bullard proposed several different distributions of density, there were many possible Earth-models.

The numerical values of these hypothetical distributions were tabulated at given radii. For the calculation of eigenfrequencies, numerical values at radial steps are needed, and since the number of such steps is large, values of the material constants at intermediate points must be found by interpolation from the original tabulated values. Different workers have used different methods of interpolation and solution, and also different radial steps. Thus, even if the same model is used, different workers will probably obtain different sets of eigenvalues, and it is almost impossible to compare these in order to identify the best method of numerical computation or the best set of computed modes.

Table 6.2. *Periods of observed and theoretical spheroidal modes in minutes. Fundamental mode* ($n = 0$).

	Observed period		Theoretical period			
Mode l	1964 Alaskan Earthquake	Mean of Chilean data (Press, 1964)	Jeffreys–Bullen A (Alsop, 1963)	Jeffreys–Bullen B (Alsop, 1963)	Gutenberg–Bullen A (Alsop, 1963)	UTD 124A′ (Dziewonski & Gilbert, 1972)
2	54.14	53.95	53.43	53.72	53.50	53.80
3	35.40	35.62	35.25	35.52	35.32	35.53
4	–	25.82	25.50	25.75	25.53	25.74
5	–	19.86	19.62	19.85	19.65	19.82
6	16.15	16.07	15.90	16.17	15.92	16.05
7	13.53	13.52	13.42	13.65	13.43	13.54
8	11.80	11.76	11.73	11.97	11.73	11.80
9	10.58	10.57	10.53	10.77	10.53	10.57
10	9.659	9.629	9.650	9.883	9.650	9.663
11	8.937	8.917	8.950	9.183	8.950	8.958
12	8.368	8.362	8.383	8.617	8.383	8.382
13	7.867	7.875	7.887	8.105	7.887	7.895
14	7.484	7.465	7.473	7.680	7.475	7.476
15	7.136	7.102	7.107	7.300	7.108	7.109
16	6.768	6.778	6.780	6.962	6.782	6.786
17	6.482	6.492	6.488	6.658	6.492	6.498
18	6.262	6.231	6.227	6.383	6.230	6.240
19	6.178	5.998	5.988	6.133	5.993	6.007
20	5.901	5.778	5.772	5.907	5.778	5.795
21	5.613	5.595	5.575	5.698	5.582	5.602
22	5.415	5.410	5.392	5.505	5.402	5.425
23	5.230	5.256	5.225	5.327	5.235	5.258
24	5.171	5.105	5.068	5.162	5.080	5.106
25	4.949	4.959	4.923	5.007	4.937	4.963
26	4.819	4.827	4.787	4.862	4.802	4.830
27	4.720	4.694	4.660	4.727	4.675	4.705
28	4.557	4.588	4.540	4.600	4.557	4.588
29	4.490	4.469	4.427	4.478	4.445	4.476

Tables 6.1 and 6.2, compare the results of such calculations with observed values (Nowroozi, 1965) for fundamental modes of toroidal and spheroidal oscillations; corresponding results of Dziewonski & Gilbert (1972) are given in the last column of each table. Toroidal and spheroidal eigenfrequencies of the first three overtones for $l \leq 5$, given by Dziewonski & Gilbert (1972) are shown in Table 6.3.

Table 6.3. *Periods of toroidal and spheroidal modes in minutes for* UTD 124A′ *Earth-model for* $l = 1$ *to* 5, $n = 0$ *to* 3.

Modes		Toroidal		Spheroidal	
l	n	Theoretical	Observed	Theoretical	Observed
1	1	13.45		41.30	
	2	7.635		17.63	17.63
	3	5.216		11.79	
2	0	43.82	44.01	53.80	53.89
	1	12.60	12.61	24.50	24.51
	2	7.487		15.07	15.07
	3	5.166		9.680	9.680
3	0	28.35	28.43	35.53	35.56
	1	11.56	11.59	17.73	17.68
	2	7.278		13.43	13.40
	3	5.093		8.142	8.151
4	0	21.75	21.92	25.74	25.79
	1	10.48	10.50	14.20	14.21
	2	7.017	7.030	12.10	12.08
	3	5.000		7.311	7.319
5	0	17.92	17.93	19.82	19.84
	1	9.512		12.17	12.18
	2	6.719		11.01	11.01
	3	4.889		6.916	6.919

In Table 6.4† we list eigenperiods of the gravest modes of spheroidal and toroidal oscillations, calculated from various Earth-models. Comparison of calculated values with observed values in Table 6.4, with its multiplicity of models and methods, shows the desirability of introducing a common reference Earth-model. As described in Chapters 1 and 10, the Standard Earth-Model Committee of the International Union of Geodesy and Geophysics has been entrusted with the task of recommending a reference model of the Earth which subsequent investigators can use as starting point, and to which proposed corrections or improvements may be referred.

†This table is not a complete list of all published values.

Table 6.4. *Eigenperiod of the gravest mode (in minutes).*

	Name of Earth-model	Spheroidal ($l = 2, n = 0$)	Toroidal ($l = 2, n = 0$)	Reference†
THEORETICAL	Jeffreys–Bullen A	53.43		Alsop (1963)
	Jeffreys–Bullen A		43.50	L
	Jeffreys–Bullen A′	53.46	43.51	L
	Jeffreys–Bullen B	53.71 ($\mu_c = 0$)	44.24	L
	Jeffreys–Bullen B	53.84		L (solid inner core)
	Jeffreys–Bullen B	53.72		Alsop (1963)
	Jeffreys–Bullen B	53.70	44.16	Pekeris *et al.* (1961 b)
	Jeffreys–Bullen		45.53	Gilbert & MacDonald (1960)
	Gutenberg–Bullen A	53.50		Alsop (1963)
	Gutenberg–Bullen A	53.52	43.61	Pekeris *et al.* (1961 b)
	Gutenberg–Bullen A′	53.54	43.61	L
	Gutenberg–Bullen B	53.78		Alsop (1963)
	Gutenberg		44.19	Gilbert & MacDonald (1960)
	Bullen		43.43	Takeuchi (1959)
	Gutenberg–Bullard I	53.89	44.31	L
	Gutenberg–Bullard II	53.19	43.14	L
	Gutenberg–Bullard III		43.72	L
	Lehman–Bullen A		43.50	L
	Modified Lehman–Bullard IV	53.89		Backus & Gilbert (1961)
	M1	53.82	43.77	L
	M3	53.78	43.75	L
	UTD 124A′	53.84	43.82	Dziewonski & Gilbert (1972)
	UTD 124B′	53.81	43.84	Dziewonski & Gilbert (1972)
OBSERVED	Chilean earthquake	53.95		Press (1964)
	Alaskan earthquake	54.14		Nowroozi (1965)
	Mean of all available data	53.89	44.01	Derr (1969a)

†L indicates Landisman, Satô & Nafe (1965)

7

EFFECT OF THE EARTH'S ROTATION ON THE EIGENFREQUENCIES OF FREE OSCILLATION

7.1 Equations of motion of an elastic sphere when the effect of rotation is included

The Earth rotates about its axis once a day, so that the angular speed of rotation Ω is $2\pi/1440\ \text{min}^{-1}$. The period of the gravest mode of free oscillation of the Earth is about 53.7 min, so that the angular frequency ω is $2\pi/53.7\ \text{min}^{-1}$. Thus the highest value of Ω/ω is about 1/27, which may be regarded as a small quantity for our present purposes.

The effects of rotation enter as body force **F** in (5.1.1) in the form

$$-2\rho\mathbf{\Omega}\wedge\partial_t\mathbf{u}-\rho\mathbf{\Omega}\wedge(\mathbf{\Omega}\wedge\mathbf{r}),\tag{7.1.1}$$

where $\mathbf{\Omega}$ is the rotation vector of the Earth, **u** the usual local displacement vector and **r** the position vector. If we take rectangular Cartesian axes $Oxyz$ with the Earth's centre as origin, and its polar axis as Oz, then $\mathbf{\Omega}$ is

$$\Omega\hat{\mathbf{z}},\tag{7.1.2}$$

where $\hat{\mathbf{z}}$ is the unit vector in the z-direction. Let S be a frame of orthogonal axes in directions defined by the polar coordinate curves at **r**. Then components of $\mathbf{\Omega}$ with regard to S are

$$(\Omega\cos\vartheta,\ -\Omega\sin\vartheta,\ 0)\tag{7.1.3}$$

The second term of (7.1.1) represents the *centrifugal force*, a force which is perpendicular to and directed away from the axis of spin; its magnitude is

$$\rho r\sin\vartheta\Omega^2,\qquad\text{where } r=|\mathbf{r}|.\tag{7.1.4}$$

The first term of (7.1.1) represents the *Coriolis force*: its components with regard to the same frame S are

$$(2\Omega\rho\sin\vartheta\partial_t w, \quad 2\Omega\rho\cos\vartheta\partial_t w, \quad -2\rho\Omega(\cos\vartheta\partial_t v+\sin\vartheta\partial_t u)), \tag{7.1.5}$$

where $\partial_t u, \partial_t v$ and $\partial_t w$ are velocity components with regard to S.

Let us define stress-gradients

$$\left.\begin{aligned}
\mathscr{K} &\equiv \rho_0 g_0\Delta+\rho_0\partial_r\psi'-\rho_0\partial_r(ug_0)+\partial_r\widehat{rr}+\frac{1}{r}\partial_\vartheta\widehat{r\vartheta}+\frac{1}{r\sin\vartheta}\partial_\varphi\widehat{r\varphi}\\
&\quad+\frac{1}{r}(2\widehat{rr}-\widehat{\vartheta\vartheta}-\widehat{\varphi\varphi}+\widehat{r\vartheta}\cot\vartheta),\\
\mathscr{L} &\equiv \frac{1}{r}\rho_0\partial_\vartheta\psi'-\frac{1}{r}\rho_0 g_0\partial_\vartheta u+\partial_r\widehat{r\vartheta}+\frac{1}{r}\partial_\vartheta\widehat{\vartheta\vartheta}+\frac{1}{r\sin\vartheta}\partial_\varphi\widehat{\vartheta\varphi}\\
&\quad+\frac{1}{r}\{(\widehat{\vartheta\vartheta}-\widehat{\varphi\varphi})\cot\vartheta+3\widehat{r\vartheta}\},\\
\mathscr{M} &\equiv \frac{1}{r\sin\vartheta}\rho_0\partial_\varphi\psi'-\frac{1}{r\sin\vartheta}\rho_0 g_0\partial_\varphi u+\partial_r\widehat{r\varphi}+\frac{1}{r}\partial_\vartheta\widehat{\vartheta\varphi}\\
&\quad+\frac{1}{r\sin\vartheta}\partial_\varphi\widehat{\varphi\varphi}+\frac{1}{r}(3\widehat{r\varphi}+2\widehat{\vartheta\varphi}\cot\vartheta).
\end{aligned}\right\} \tag{7.1.6}$$

$\mathscr{K}$, $\mathscr{L}$ and $\mathscr{M}$ form the left-hand sides of (5.1.1) for a self-gravitating and non-rotating Earth-model (cf. (5.1.12)). Using these quantities, the equations of motion of a self-gravitating, *rotating* elastic Earth-model may be written

$$\rho\partial_t^2\begin{Bmatrix}u\\ v\\ w\end{Bmatrix}+2\Omega\rho\begin{Bmatrix}-\sin\vartheta\partial_t w\\ -\cos\vartheta\partial_t w\\ \cos\vartheta\partial_t v+\sin\vartheta\partial_t u\end{Bmatrix}-\rho\Omega^2 r\sin\vartheta\begin{Bmatrix}\sin\vartheta\\ \cos\vartheta\\ 0\end{Bmatrix}=\begin{Bmatrix}\mathscr{K}\\ \mathscr{L}\\ \mathscr{M}\end{Bmatrix}. \tag{7.1.7}$$

Poisson's equation for ψ' is

$$\nabla^2\psi'=4\pi G(\rho_0\Delta+u\partial_r\rho_0), \tag{7.1.8}$$

as before: it is not affected by rotation. Let the solutions of (7.1.7) when $\Omega=0$, that is, for a non-rotating Earth-model, be denoted by subscript N. We employ a first-order perturbation calculation in the small parameter α defined by

$$\alpha = \Omega/\omega_N (\ll 1). \tag{7.1.9}$$

Since the centrifugal term is of order α^2 we neglect it in our first-order theory (but see § 7.2). If we put

$$\left.\begin{aligned} \omega &= \omega_N + \alpha\omega_1, & u &= u_N + \alpha u_1, & \mathscr{K} &= \mathscr{K}_N + \alpha\mathscr{K}_1, \\ \psi' &= \psi'_N + \alpha\psi'_1, & v &= v_N + \alpha v_1, & \mathscr{L} &= \mathscr{L}_N + \alpha\mathscr{L}_1, \\ & & w &= w_N + \alpha w_1, & \mathscr{M} &= \mathscr{M}_N + \alpha\mathscr{M}_1, \end{aligned}\right\} \tag{7.1.10}$$

and assume a time factor exp $i\omega t$ in each displacement component and in the gravitational potential ψ', the zero-order equations in α, that is, the equations (5.1.12), are written as

$$\left.\begin{aligned} \rho_0\omega_N^2 u_N + \mathscr{K}_N &= 0, \\ \rho_0\omega_N^2 v_N + \mathscr{L}_N &= 0, \\ \rho_0\omega_N^2 w_N + \mathscr{M}_N &= 0, \\ \nabla^2\psi'_N - 4\pi G(\rho_0\Delta_N + u_N\partial_r\rho_0) &= 0. \end{aligned}\right\} \tag{7.1.11}$$

Inserting (7.1.10) into (7.1.7), using (7.1.11), and putting the coefficient of α equal to zero, we obtain the equations for the unknown functions u_1, v_1, w_1 and ψ'_1:

$$\left.\begin{aligned} \rho\omega_N^2 u_1 + \mathscr{K}_1 &= -2\rho\omega_N\omega_1 u_N - 2i\rho\omega_N^2 w_N \sin\vartheta, \\ \rho\omega_N^2 v_1 + \mathscr{L}_1 &= -2\rho\omega_N\omega_1 v_N - 2i\rho\omega_N^2 w_N \cos\vartheta, \\ \rho\omega_N^2 w_1 + \mathscr{M}_1 &= -2\rho\omega_N\omega_1 w_N \\ &\quad + 2i\rho\omega_N^2(v_N\cos\vartheta + u_N\sin\vartheta), \\ \nabla^2\psi'_1 &= 4\pi G(\rho\Delta_1 + u_1\partial_r\rho_0). \end{aligned}\right\} \tag{7.1.12}$$

The first three equations of (7.1.12) are, in vector form,

$$\rho\omega_N^2\mathbf{u}_1 + \begin{Bmatrix} \mathscr{K}_1 \\ \mathscr{L}_1 \\ \mathscr{M}_1 \end{Bmatrix} = -2\rho\omega_N\omega_1\mathbf{u}_N - 2i\,\rho\omega_N^2\mathbf{u}_N \wedge \hat{\mathbf{z}}. \tag{7.1.13}$$

We now confine ourselves to the evaluation of ω_1, the deviation of eigenfrequency from that for a non-rotating sphere. Let an asterisk denote the complex conjugate. Then we have from (7.1.13), multiplying by $\mathbf{u}_\mathrm{N}^*$ and integrating over the volume V of the sphere,

$$\int_V \rho\omega_\mathrm{N}^2 \mathbf{u}_1 \cdot \mathbf{u}_\mathrm{N}^* \mathrm{d}V + \int_V \begin{Bmatrix} \mathscr{K}_1 \\ \mathscr{L}_1 \\ \mathscr{M}_1 \end{Bmatrix} \cdot \mathbf{u}_\mathrm{N}^* \mathrm{d}V$$

$$= -2\omega_\mathrm{N}\omega_1 \int_V \rho \mathbf{u}_\mathrm{N} \cdot \mathbf{u}_\mathrm{N}^* \mathrm{d}V + 2\mathrm{i}\omega_\mathrm{N}^2 \int_V \rho(\mathbf{u}_\mathrm{N} \wedge \mathbf{u}_\mathrm{N}^*) \cdot \hat{\mathbf{z}} \mathrm{d}V. \quad (7.1.14)$$

Solutions of (7.1.7) for a non-rotating sphere are given in (5.2.2) and (5.2.3). The equations of motion are linear in the unknown functions u, v, w and ψ'. $u_\mathrm{N} + \alpha u_1$, $v_\mathrm{N} + \alpha v_1$, $w_\mathrm{N} + \alpha w_1$ and $\psi'_\mathrm{N} + \alpha\psi'_1$ must satisfy the same boundary conditions as u_N, v_N, w_N and ψ'_N, for all values of ϑ and φ, so we may reasonably assume that u_1, v_1, w_1 and ψ'_1 have similar forms to those of u_N, v_N, w_N and ψ'_N, namely:

$$\left.\begin{aligned} u_1 &= U_1(r) P_l^m(\cos\vartheta) \exp im\varphi, \\ v_1 &= V_1(r) \partial_\vartheta P_l^m(\cos\vartheta) \exp im\varphi, \\ w_1 &= imV_1(r)(1/\sin\vartheta) P_l^m(\cos\vartheta) \exp im\varphi, \\ \psi'_1 &= Y_1(r) P_l^m(\cos\vartheta) \exp im\varphi. \end{aligned}\right\} \quad (7.1.15)$$

Exp $im\varphi$ is used here, instead of $\genfrac{}{}{0pt}{}{\cos}{\sin} m\varphi$ as in (5.2.2) and (5.2.3), to simplify the manipulation of complex conjugates. The expressions (7.1.15) represent a solution of (7.1.13) for any integral l and m, provided $-l \leq m \leq l$.

The left-hand side of (7.1.14) is the sum of the first variation of kinetic energy and potential energy due to variation $\mathbf{u}_1$ from $\mathbf{u}_\mathrm{N}$ when, the frequency ω does not vary and Ω is zero. It can be proved, after considerable manipulation (Pekeris *et al.*, 1961a), using the fourth equation of (7.1.12) and the imposed boundary conditions, that the sum of the two integrals on the left-hand side of (7.1.14) vanishes. Then, equating to zero the integral on the right-hand side of (7.1.14), we have

$$2\omega_\mathrm{N}\omega_1 \int_V \rho[U_\mathrm{N}^2(P_l^m)^2 + V_\mathrm{N}^2\{(\partial_\vartheta P_l^m)^2 + m^2(P_l^m)^2/\sin^2\vartheta\}]\mathrm{d}V$$

$$= 4m\omega_\mathrm{N}^2 \int_V \rho\{U_\mathrm{N} V_\mathrm{N}(P_l^m)^2 + V_\mathrm{N}^2 \cot\vartheta P_l^m \partial_\vartheta P_l^m\}\mathrm{d}V. \quad (7.1.16)$$

Using the standard formulae (Pekeris *et al.*, 1961a)

$$\left.\begin{aligned}
&\int_0^\pi \sin\vartheta (P_l^m)^2\,\mathrm{d}\vartheta = A, && A = \frac{2}{2l+1}\frac{(l+m)!}{(l-m)!},\\
&\int_0^\pi \operatorname{cosec}\vartheta (P_l^m)^2\,\mathrm{d}\vartheta = C, && C = \frac{1}{m}\frac{(l+m)!}{(l-m)!},\\
&\int_0^\pi \sin\vartheta (\partial_\vartheta P_l^m)^2\,\mathrm{d}\vartheta = l(l+1)A - m^2 C, &&\\
&\int_0^\pi \cos\vartheta P_l^m \partial_\vartheta P_l^m\,\mathrm{d}\vartheta = \tfrac{1}{2}A, &&
\end{aligned}\right\} \quad (7.1.17)$$

we obtain from (7.1.16)

$$\omega_1 = m\omega_{\mathrm{N}}\tau(l),$$

where

$$\tau(l) = \int_0^a \rho r^2 (2U_{\mathrm{N}}V_{\mathrm{N}} + V_{\mathrm{N}}^2)\mathrm{d}r \bigg/ \int_0^a \rho r^2\{U_{\mathrm{N}}^2 + l(l+1)V_{\mathrm{N}}^2\}\,\mathrm{d}r, \quad (7.1.18)$$

and

$$m = -l, -(l-1), \dots, 0, \dots, (l-1), l. \quad (7.1.19)$$

From (7.1.14) we see that $\omega_{\mathrm{N}}\tau(l)$ is proportional to the ratio of the rate of working of the Coriolis force to the average kinetic energy of the mode when there is no rotation. For toroidal oscillations the radial displacement is zero. Then U_{N} disappears and these two quantities have the same form as functions of r, and we obtain from (7.1.18)

$$\tau(l) = \int_0^a \rho r^2 V_{\mathrm{N}}^2\mathrm{d}r \bigg/ l(l+1)\int_0^a \rho r^2 V_{\mathrm{N}}^2\mathrm{d}r = 1/l(l+1). \quad (7.1.20)$$

Thus it has been shown that the effect of rotation of the Earth is to split each eigenfrequency of Legendre degree l into $2l+1$ frequencies

$$\omega = \omega_{\mathrm{N}} + m\Omega\tau(l), \qquad -l \le m \le l. \quad (7.1.21)$$

This phenomenon is called 'rotational splitting'. The degeneration of modes for a non-gravitating, homogeneous and isotropic elastic sphere, which was described in § 2.3, is released by rotation of the sphere.

Backus & Gilbert have shown as follows (Backus & Gilbert, 1961) that for given l the geographical amplitude pattern is that of a standing wave that travels slowly westward.

Consider motion *symmetric* about a source at (ϑ_0, φ_0) on the surface of a non-rotating sphere. The angular dependence of the displacement will be $P_l(\cos\Theta)$, where Θ is the angle between radii to source and observing station (ϑ, φ). Thus the local movement in a mode of angular frequency ω_{N} will be proportional to

$$P_l\{\cos\vartheta\cos\vartheta_0 + \sin\vartheta\sin\vartheta_0\cos(\varphi-\varphi_0)\}\sin\omega_{\mathrm{N}}t. \quad (7.1.22)$$

By the addition theorem for Legendre functions, this is proportional to

$$\mathrm{Im}\sum_{m=-l}^{l}\frac{(l-m)!}{(l+m)!}P_l^m(\cos\vartheta)P_l^m(\cos\vartheta_0)$$
$$\times\exp\{im(\varphi-\varphi_0)+\mathrm{i}\omega_{\mathrm{N}}t\}. \quad (7.1.23)$$

If we now introduce rotation Ω, so that ω_{N} is replaced, as in (7.1.21), by

$$\omega_{\mathrm{N}} + m\Omega\tau(l)$$

in the split mode of order m, (7.1.23) becomes

$$\mathrm{Im}\sum_{m=-l}^{l}\frac{(l-m)!}{(l+m)!}P_l^m(\cos\vartheta)P_l^m(\cos\vartheta_0)$$
$$\times\exp\{im(\varphi-\varphi_0)+im\Omega\tau t+\mathrm{i}\omega_{\mathrm{N}}t\}. \quad (7.1.24)$$

Thus the angular pattern drifts with speed $\Omega\tau$ westward.

If we consider the standing wave (7.1.22) for the non-rotating sphere as the result of superposition of equal east-going ($m<0$) and west-going ($m>0$) waves, the former is retarded and the latter advanced by the rotation Ω; the standing wave consequently travels steadily westward across the meridians.

7.2 Theoretical and observed values of rotational splitting

The result obtained in the first-order perturbation calculation of the previous section is not sufficient for comparison with observation.

Dahlen (1968, 1969) has developed a second-order perturbation theory, taking the effects of ellipticity and transverse heterogeneity as well as that of rotation into consideration. According to second-order theory (Dahlen, 1968) the eigenfrequency of a mode split by rotation and ellipticity is given by

$$\omega/\omega_{\mathrm{N}} = 1 + \alpha(n, l) + m\beta(n, l) + m^2\gamma(n, l), \tag{7.2.1}$$

where

$$\left.\begin{aligned} \alpha(n, l) &= \alpha_{\mathrm{r}}(n, l)(\Omega/\omega_{\mathrm{N}})^2 + \alpha_{\mathrm{e}}(n, l)\varepsilon, \\ \beta(n, l) &= \beta_{\mathrm{r}}(n, l)(\Omega/\omega_{\mathrm{N}}), \\ \gamma(n, l) &= \gamma_{\mathrm{r}}(n, l)(\Omega/\omega_{\mathrm{N}})^2 + \gamma_{\mathrm{e}}(n, l)\varepsilon, \end{aligned}\right\} \tag{7.2.2}$$

n is (radial) overtone number, l Legendre degree of the mode, and subscripts r and e indicate the effects of rotation and ellipticity; ε is the ellipticity of the Earth, which is about 1/300. $\beta_{\mathrm{r}}(n, l)$ coincides with $\tau(l)$ of (7.1.18). Thus the second-order effect of rotation is to shift each of the split frequencies by an amount $\alpha + m^2\gamma$ (which is non-zero even for $m = 0$), thus changing the frequency interval between adjacent split components. The effect of ellipticity is like the second-order rotation effect, and of comparable magnitude. It shifts each split frequency by $(\alpha_{\mathrm{e}} + \gamma_{\mathrm{e}} m^2)\varepsilon$.

Dahlen showed further (Dahlen, 1969, Luh, 1973) that the perturbed ${}_nS_l^m$ mode will have contributions due to $S_{l\pm 2}^m$ and $T_{l\pm 1}^m$, while the perturbed ${}_nT_l^m$ mode will have contributions due to $T_{l\pm 2}^m$ and $S_{l\,\pm 1}^m$. This coupling may be strong if the frequencies of the coupled modes are close. It is clear that m must be the same for any pair of coupled modes, since the effects of rotation and ellipticity are independent of longitude. It was noted by Dahlen that coupling must be expected between ${}_0S_l$ and ${}_0T_{l+1}$ in the range $l = 10, 11, \ldots,$ 22 where the eigenfrequencies run close (Dahlen, 1969).

Values of $\tau(l)$ for spheroidal oscillations are given by Pekeris *et al.* (1961a), and Dahlen (1968) who used an SNREI model (spherical, non-rotating, elastic, isotropic) as starting point:

$l = 2$	$n = 0$	$\tau(2) = 0.395$	(Gutenberg model, Pekeris *et al.*)
		$= 0.395$	(Bullen B model, Pekeris *et al.*)
		$= 0.3994$	(SNREI model I, Dahlen)

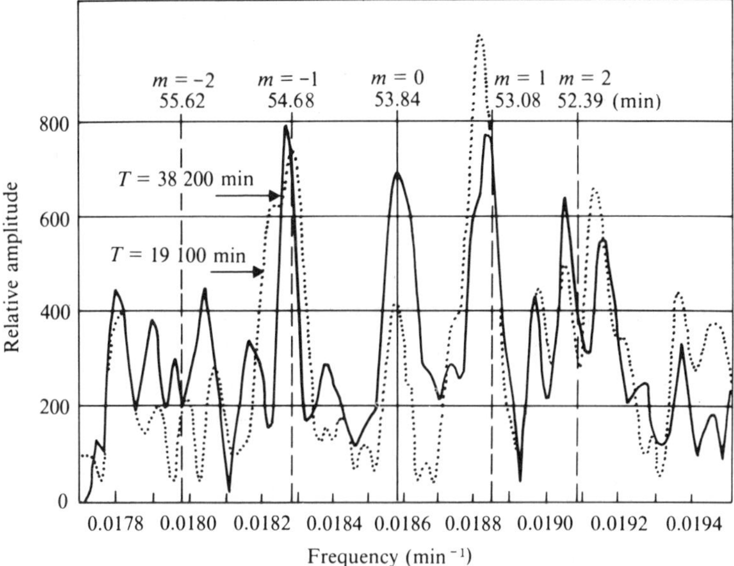

Fig. 7.1. Fourier spectrum of a record of the Chilean earthquake of 1960 on a quartz strain gauge at Isabella, California. Segment of Fourier analysis of two lengths of record, 19 100 min and 38 200 min, sample interval 2 min. Dashed vertical lines represent computed positions of peaks corresponding to $m = \pm 1, \pm 2$ relative to $m = 0$.

$$
\begin{aligned}
l = 3 \quad n = 0 \qquad \tau(3) &= 0.183 \quad \text{(Gutenberg model, Pekeris \textit{et al.})}\\
&= 0.182 \quad \text{(Bullen B model, Pekeris \textit{et al.})}\\
&= 0.1868 \quad \text{(SNREI model I, Dahlen)}
\end{aligned}
$$

The Chilean earthquake of 22 May 1960 ($M = 8.5$) excited various modes of free oscillation of the Earth, providing good data for the study of rotational splitting. Fig. 7.1 shows the Fourier spectrum of a record of the spheroidal mode $l = 2$, $n = 0$ from the strain-meter at Isabella, California. Fourier analysis was done on two lengths of record, 19 100 min and 38 200 min, with sample intervals 2 min (Smith, 1961, Dahlen, 1968). In Fig. 7.1 the position assumed for the central line $m = 0$ is indicated by the solid vertical line; the relative computed theoretical positions (using SNREI model I above and (7.2.2)) of the other four lines corresponding to $m = \pm 1, m = \pm 2$ are indicated by dashed vertical lines. The agreement between theory and observation is seen to be fairly satisfactory.

The analysis of the spheroidal mode $l = 3$ is given as:

$l = 3$	$m = -3$	$T = 35.99$ min	
	$= -2$	$= 35.83$	$T_g = 35.87$ min $T_s = 35.9$ min
	$= -1$	$= 35.67$	
	$= 0$	$= 35.50$	
	$= 1$	$= 35.34$	
	$= 2$	$= 35.18$	$T_g = 35.24$ min $T_s = 35.2$ min
	$= 3$	$= 35.03$	

T is the eigenperiod calculated by Pekeris *et al.* (1961a) for a Bullen B model using first-order perturbation theory (§ 7.1). T_g and T_s are periods observed by gravimeter and strain-meter, respectively. Pekeris and his colleagues also calculated spectral amplitudes and obtained satisfactory agreement with observation.

Gravimeter data from the Alaskan earthquake of 27 March 1964 ($M = 8.4$) also show rotational splitting of the spheroidal mode $l = 2$ (Slichter, 1967).

Rotational splitting of the toroidal modes has not yet been observed. This may be due to attenuation of toroidal oscillations, which would be accompanied by broadening of peaks in the spectrum. Attenuation is more severe for toroidal than spheroidal oscillations, since shearing motions are more strongly damped in the Earth than compressional motions, as is seen from the comparative decay of S and P waves with distance.

8

ASYMPTOTIC DISTRIBUTION OF EIGENFREQUENCIES

In 1974 Anderssen & Cleary (1974) set out to discover whether the frequencies of radial overtones in free toroidal oscillations of the Earth behaved asymptotically, for large overtone number, in the way that would be expected of eigenvalues of a *Sturm–Liouville* equation – to which category the equation (6.1.19) for $W(r)$ belongs. By numerical studies of four Earth-models they were able to show that, while in general the pattern of frequencies had the expected asymptotic behaviour, there were deviations which seemed to depend on the severity of the discontinuities of elastic parameters within the model. (See also Cleary, Osborne & Anderssen, 1974).

In this chapter we first investigate the asymptotic behaviour of eigenfrequencies for *smooth* Earth-models, following the method of Gilbert (1975), and then examine the effect of discontinuities.

8.1 Asymptotic distribution of radial eigenfrequencies for smooth Earth-models

Throughout this chapter we neglect gravity and rotation. In this section we examine sequences of eigenfrequencies for smooth Earth-models, in order of increasing complexity – first purely radial modes ${}_nS_0$, then toroidal modes ${}_nT_l$, and finally spheroidal modes ${}_nS_l$.

8.1.1 *Purely radial modes*

We first consider purely radial vibrations, in which $l=0, m=0$, $\partial/\partial\vartheta \equiv 0, \partial/\partial\varphi \equiv 0$. This is the type ${}_nS_0$. Then from (2.3.5), using a dot to denote a derivative with regard to r as before,

$$\left.\begin{aligned} e_{rr} &= \dot{u}, & e_{r\vartheta} &= 0, \\ e_{\vartheta\vartheta} &= u/r, & e_{r\varphi} &= 0, \\ e_{\varphi\varphi} &= u/r, & e_{\vartheta\varphi} &= 0, \end{aligned}\right\} \tag{8.1.1}$$

and so $\Delta = \dot{u} + 2u/r$. Then from (5.1.10)

$$\left.\begin{aligned} \widehat{rr} &= \lambda\Delta + 2\mu\dot{u} = (\lambda + 2\mu)\dot{u} + 2\lambda u/r, \\ \widehat{\vartheta\vartheta} &= \widehat{\varphi\varphi} = \lambda\Delta + 2\mu u/r, \\ \widehat{r\vartheta} &= 0,\ \widehat{r\varphi} = 0,\ \widehat{\vartheta\varphi} = 0. \end{aligned}\right\} \tag{8.1.2}$$

Hence the first equation of (5.1.1) is

$$\partial_r(\lambda\Delta) + 2\partial_r(\mu\partial_r u) + (\mu/r)(4\partial_r u - 4u/r) = \rho\partial_t^2 u, \tag{8.1.3}$$

while the second and third equations vanish identically. For a harmonic motion of angular frequency ω, the right-hand side of (8.1.3) becomes $-\rho\omega^2 u$.

We now introduce new variables y_i as in (6.1.26); but here $i = 1, 2$ only. y_1 is the radial displacement u and y_2 the stress-component $\widehat{rr}$. Then from (6.1.27) we obtain

$$\dot{y}_1 = -\frac{2\lambda}{\lambda + 2\mu}\frac{1}{r}y_1 + \frac{1}{\lambda + 2\mu}y_2, \tag{8.1.4}$$

$$\dot{y}_2 = \left\{-\rho\omega^2 + \frac{4\mu(3\lambda + 2\mu)}{\lambda + 2\mu}\frac{1}{r^2}\right\}y_1 - \frac{4\mu}{\lambda + 2\mu}\frac{1}{r}y_2. \tag{8.1.5}$$

We can find the *form* of the solution for large ω by taking λ, μ and ρ constant, independent of r, and assuming that y_1/r^2, $y_2/(\lambda + 2\mu)r$, $\dot{y}_1/r$ are small compared with $\ddot{y}_1$ and $\omega^2 y_1$. Then neglecting in (8.1.5) all terms of order lower than $\omega^2 y_1$, and using (8.1.4), we find

$$y_1 \simeq A\sin(\omega r/c_{\mathrm{P}} + \phi), \qquad \text{where } c_{\mathrm{P}}^2 = (\lambda + 2\mu)/\rho, \tag{8.1.6}$$

and A and ϕ are independent of r, determined by boundary conditions. Then

$$y_2 \simeq (\lambda + 2\mu)\frac{\omega}{c_{\mathrm{P}}}A\cos(\omega r/c_{\mathrm{P}} + \phi), \tag{8.1.7}$$

which is $O(\omega y_1)$. We may now check that our assumptions concerning y_1, y_2, $\dot{y}_1$ were valid.

In place of (y_1, y_2), which are of different orders in ω, let us introduce $U = y_1$ and $R = y_2/\omega$. Then reverting to the original problem, with non-uniform λ, μ, ρ, and keeping only terms $O(\omega)$ in (8.1.4) and $O(\omega^2)$ in (8.1.5), we have

$$\dot{U} = \{\omega/(\lambda + 2\mu)\}R, \tag{8.1.8}$$

$$\omega \dot{R} = -\rho \omega^2 U, \tag{8.1.9}$$

that is

$$\ddot{U} = -\frac{\rho}{\lambda + 2\mu}\omega^2 U + \omega R \frac{\mathrm{d}}{\mathrm{d}r}\left(\frac{1}{\lambda + 2\mu}\right). \tag{8.1.10}$$

But the last term in (8.1.10) is $O(1/\omega)$ times the other two as long as $\lambda + 2\mu$ does not change sharply. So we have to solve

$$\ddot{U} + \frac{\omega^2}{\{c_{\mathrm{P}}(r)\}^2} U = 0. \tag{8.1.11}$$

To do this we change the independent variable by Liouville's substitution:

$$s = \frac{1}{\gamma_{\mathrm{P}}}\int_0^r \frac{\mathrm{d}r}{c_{\mathrm{P}}}, \qquad \frac{\mathrm{d}}{\mathrm{d}r} = \frac{1}{c_{\mathrm{P}}\gamma_{\mathrm{P}}}\frac{\mathrm{d}}{\mathrm{d}s}, \tag{8.1.12}$$

where

$$\gamma_{\mathrm{P}} = \int_0^a \frac{\mathrm{d}r}{c_{\mathrm{P}}}, \tag{8.1.13}$$

which is the time required for a P wave to traverse the radius of the sphere. We obtain, in place of (8.1.11), with $U = c_{\mathrm{P}}^{1/2}\omega^{-1/2} V$,

$$V'' + \omega^2\gamma_{\mathrm{P}}^2 V + \left(\tfrac{1}{2}c_{\mathrm{P}}^{-1}c_{\mathrm{P}}'' - \tfrac{3}{4}c_{\mathrm{P}}^{-2}c_{\mathrm{P}}'^2\right)V = 0, \tag{8.1.14}$$

where primes indicate derivatives with regard to s. Provided c_{P}' and c_{P}'' are not large compared with c_{P}, (8.1.14) has the asymptotic solution

$$V \simeq B \sin(\omega\gamma_{\mathrm{P}} s + \psi), \tag{8.1.15}$$

where B and ψ are adjustable constants.

Then

$$R \simeq (\lambda + 2\mu)c_{\mathrm{P}}^{-1/2}\omega^{-1/2} B \cos(\omega\gamma_{\mathrm{P}} s + \psi). \tag{8.1.16}$$

The boundary condition is $R = 0$ when $r = a$, that is, when $s = 1$. Thus we must have $\cos(\omega\gamma_{\mathrm{P}} + \psi) = 0$. The roots of this equation in ω are $\omega_n = \{(n + \tfrac{1}{2})\pi - \psi\}/\gamma_{\mathrm{P}}$, and have separation

$$\pi/\gamma_{\mathrm{P}}. \tag{8.1.17}$$

For large $n, \omega_n \simeq n\pi/\gamma_{\mathrm{P}}$. Thus the asymptotic behaviour of the eigenfrequencies of overtones of type ${}_nS_0$ is controlled by $\gamma_P = \int_0^a dr/c_P(r)$, interpreted as above (8.1.13).

8.1.2 *Toroidal modes*

In toroidal modes, gravity has no effect, and the following working is valid (without approximation) for a self-gravitating sphere. We assume that the sphere has a radially heterogeneous mantle and a liquid core and that there is no tangential stress between mantle and core at the interface.

We seek modes symmetrical about the axis $\vartheta = 0$, and put $m = 0, \partial/\partial\varphi \equiv 0$. Then from (6.1.18)

$$u = 0, \quad v = 0, \quad w = B_l W(r)\partial_\vartheta P_l(\cos\vartheta). \tag{8.1.18}$$

Here the strains are

$$e_{rr} = 0, \qquad e_{\vartheta\vartheta} = 0, \qquad e_{\varphi\varphi} = 0, \qquad \Delta = 0,$$

$$e_{r\vartheta} = 0, \qquad e_{r\varphi} = \frac{1}{2}\left(\dot{w} - \frac{1}{r}w\right), \qquad (e_{\vartheta\varphi} \text{ is not required}),$$

and the stresses are zero except $\widehat{\vartheta\varphi}$, which we do not use, and

$$\widehat{r\varphi} = \mu\left(\dot{W} - \frac{1}{r}W\right)B_l\partial_\vartheta P_l(\cos\vartheta). \tag{8.1.19}$$

Substituting these into (5.1.1) we get, as in (6.1.19),

$$\mu\left(\ddot{W} + \frac{2}{r}\dot{W}\right) + \dot{\mu}\left(\dot{W} - \frac{1}{r}W\right) + W\left\{\rho\omega^2 - \frac{l(l+1)\mu}{r^2}\right\} = 0, \; b \le r \le a. \tag{8.1.20}$$

Zero tangential stress on inner and outer boundaries of the shell gives

$$\dot{W} - \frac{1}{r}W = 0 \qquad \text{at } r = b, a. \tag{8.1.21}$$

We put $W = rVe^{i\omega t}$, so as to obtain the standard Sturm–Liouville form:

$$\frac{d}{dr}\left(r^4\mu\frac{dV}{dr}\right) + V\{\rho\omega^2 r^4 - (l+2)(l-1)\mu r^2\} = 0,$$

with

$$\frac{dV}{dr} = 0 \qquad \text{at } r = b, \ a. \tag{8.1.22}$$

Next we introduce Liouville's transformation, with the aim of transforming the differential equation (8.1.22) into a form which can be compared, for large ω, with that for simple harmonic motion: we write

$$\left. \begin{aligned} s &= \int_b^r \frac{dr}{c_S(r)}, \qquad \frac{d}{dr} = \frac{1}{c_S}\frac{d}{ds}, \qquad (c_S = \mu^{1/2}\rho^{-1/2}), \\ M(r) &= r^2 \rho^{1/2} c_S^{1/2}, \qquad V = Z/M, \end{aligned} \right\} \tag{8.1.23}$$

and, after some reduction, get

$$Z'' + \{\omega^2 - q(s)\} Z = 0, \tag{8.1.24}$$

where

$$q(s) = \frac{M''}{M} + \frac{(l+2)(l-1)c_S^2(s)}{r^2(s)}, \tag{8.1.25}$$

and primes indicate derivatives with regard to the new independent variable s.

The boundary conditions become

$$Z'M - M'Z = 0 \tag{8.1.26}$$

at $s = 0$ and $s = \gamma_S$, where $\gamma_S = \int_b^a dr/c_S$, the time taken by a shear wave to travel radially through the shell.

If $\omega^2 \gg q(s)$ throughout the range $(0, \gamma_S)$ of s, the solution is asymptotically

$$Z \simeq C \sin(\omega s + \chi), \tag{8.1.27}$$

where C and χ are adjustable constants. The boundary conditions give a trigonometric equation for $\omega\gamma_S$, so that for large n (and provided l is not too large) the eigenfrequencies are asymptotically $n\pi/\gamma_S$, as in the previous section, except that γ_S replaces γ_P.

8.1.3 *Spheroidal modes*

Here we use the set of equations (6.1.27), except that since we neglect gravity we put $y_5 = 0, y_6 = 0$ and write $\rho(r)$ instead of $\rho_0(r)$. The physical meanings of the variables are:

$$\left. \begin{aligned} y_1 &= U, & y_2 &= \widehat{rr} = \lambda\Delta + 2\mu\dot{U}, \\ y_3 &= V, & y_4 &= \widehat{r\vartheta} = \mu\left(\dot{V} - \frac{V}{r} + \frac{U}{r}\right). \end{aligned} \right\} \tag{8.1.28}$$

Then

$$\left. \begin{aligned} \dot{y}_1 &= -\frac{2\lambda}{\lambda+2\mu}\frac{1}{r}y_1 + \frac{1}{\lambda+2\mu}y_2 + \frac{l(l+1)\lambda}{\lambda+2\mu}\frac{1}{r}y_3, \\ \dot{y}_2 &= \left\{-\rho\omega^2 + \frac{4\mu(3\lambda+2\mu)}{\lambda+2\mu}\frac{1}{r^2}\right\}y_1 - \frac{4\mu}{\lambda+2\mu}\frac{1}{r}y_2 \\ &\quad - \frac{l(l+1)2\mu(3\lambda+2\mu)}{\lambda+2\mu}\frac{1}{r^2}y_3 + \frac{l(l+1)}{r}y_4, \\ \dot{y}_3 &= -\frac{1}{r}y_1 + \frac{1}{r}y_3 + \frac{1}{\mu}y_4, \\ \dot{y}_4 &= -\frac{2\mu(3\lambda+2\mu)}{\lambda+2\mu}\frac{1}{r^2}y_1 - \frac{\lambda}{\lambda+2\mu}\frac{1}{r}y_2 \\ &\quad + \left[-\rho\omega^2 + \frac{2\mu}{\lambda+2\mu}\left\{\begin{matrix}\lambda(2l^2+2l-1) \\ +2\mu(l^2+l-1)\end{matrix}\right\}\frac{1}{r^2}\right]y_3 - \frac{3}{r}y_4. \end{aligned} \right\} \tag{8.1.29}$$

When ω^2 is very large (and provided l is not too large) these decouple, and as in § 8.1.1 the pairs reduce to:

$$\left. \begin{aligned} \dot{y}_1 &\simeq \frac{1}{\lambda+2\mu}y_2, \\ \dot{y}_2 &\simeq -\rho\omega^2 y_1, \end{aligned} \right\} \tag{8.1.30}$$

$$\left. \begin{aligned} \dot{y}_3 &\simeq \frac{1}{\mu}y_4, \\ \dot{y}_4 &\simeq -\rho\omega^2 y_3. \end{aligned} \right\} \tag{8.1.31}$$

Thus the system is decoupled into pairs of equations for $(U, \widehat{rr})$ and $(V, \widehat{r\vartheta})$. The first is identical with the pair (8.1.9) and the second with the pair which could have been obtained from (8.1.20) if we had not preferred immediate application of Liouville's transformation.

Hence there are here two separate types of asymptotic sequence of eigenfrequencies. The first is governed by the time taken by a P wave to traverse a radius of the sphere. It is called PKIKP-type by Gilbert. The second is governed by the travel-time of an S wave. This may be either in the mantle (ScS-type) or in the inner core (J-type) according to the range over which we solve the equations (8.1.31). The existence of J-type modes depends on that of an inner core of non-zero rigidity: but at present we have no way to detect J-type modes.

So far we have assumed that there are no discontinuities in ρ, λ, μ or their derivatives within the sphere. It has been shown that the use of asymptotic forms of the Sturm–Liouville equation depends critically on these assumptions (Lapwood, 1975, McNabb, Anderssen & Lapwood, 1976).

8.2 Anomalous behaviour of eigenfrequencies for realistic Earth-models

There is not at present much direct observational evidence concerning high overtones. But it is possible to examine Earth-models proposed by different workers and to calculate directly from them – that is, without appeal to asymptotic theory – the distribution of frequencies of overtones for each. This was done for toroidal oscillations by Anderssen & Cleary (1974). Table 8.1 shows periods $2\pi/{}_n\omega_l$ calculated for Haddon–Bullen HB 1 (Bullen & Haddon, 1967) and Dziewonski–Gilbert UTD 124A′ (1972). These models differ considerably, as is shown, for example, by the graphs of shear-modulus against depth (Fig. 8.1). We see in Table 8.1 the regular decrease of period as n increases for fixed l, and also as l increases for fixed n. The arrays of periods are very similar in spite of the difference between the models.

If we now introduce the approximation ${}_n\gamma_l$ to the radial travel-time for S in the mantle (γ_S of § 8.1.2), defined by ${}_n\omega_l = n\pi/{}_n\gamma_l$, we obtain Table 8.2 from Table 8.1. From this new table we see how our estimate ${}_n\gamma_l$ of γ oscillates, after $n = 0$, for both $l = 2$ and $l = 4$, in the neighbourhood of 466 for HB 1 and of 480 for UTD 124A′. For $l = 18$, the initial irregularities ($n < 10$) are much greater. Figure 8.2 shows this clearly, and shows also that ${}_n\gamma_l$ moves more smoothly with increasing n in (*a*) for HB 1 than in (*b*) for UTD 124A′. Anders-

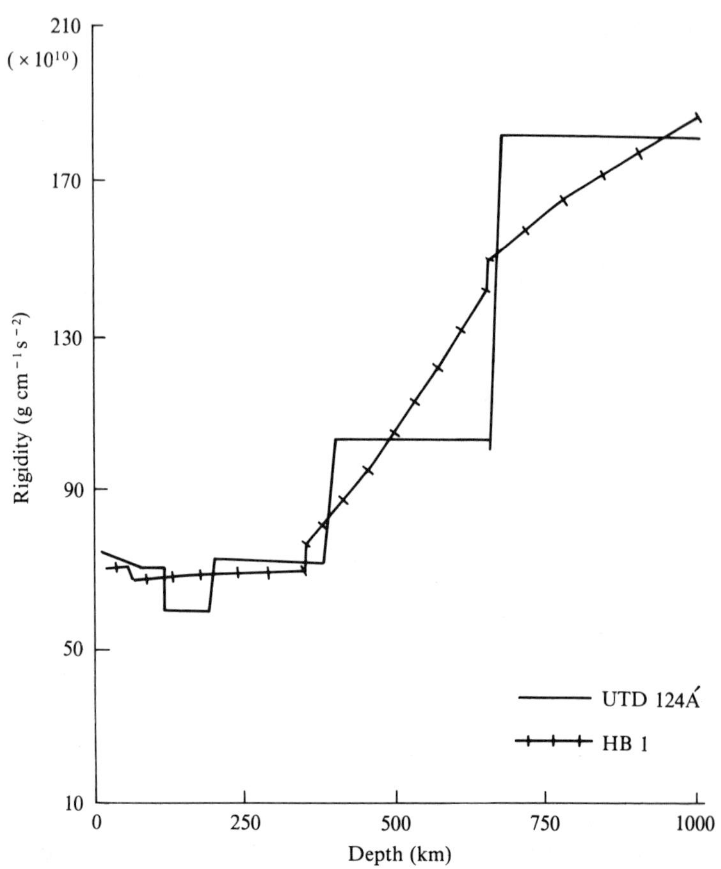

Fig. 8.1. Rigidity μ plotted against depth down to 1000 km for Earth-models HB 1 and UTD 124A′.

sen & Cleary guessed that the irregular behaviour of ${}_n\gamma_l$ for model UTD 124A′ was related to the sharper discontinuities of the model. They were able to show that by smoothing off the corners of the graph of rigidity against depth they could reduce the deviations in the graph of ${}_n\gamma_l$ against n.

It was subsequently shown (Lapwood, 1975) that the asymptotic formula considered by Anderssen & Cleary, viz.

$$ {}_n\omega_l^2 \simeq (n\pi)^2\gamma^{-2} + A\gamma^{-2} + B\gamma^{-2}n^{-2}, $$

where A and B are constants independent of n and l, is valid provided

Table 8.1

	Periods ${}_nT_l$ (s)					
	HB 1			UTD 124A′		
n	$l=2$	$l=4$	$l=18$	$l=2$	$l=4$	$l=18$
1	754.60	628.80	258.84	755.79	629.55	259.44
5	184.76	182.71	146.91	186.72	184.70	148.38
10	93.14	92.87	87.29	93.77	93.48	87.42
15	62.03	61.96	60.26	62.21	62.14	60.54
20	46.53	46.50	45.77	46.92	46.88	46.09
25	37.22	37.20	36.82	37.39	37.37	37.02
30	31.05	31.04	30.82	31.24	31.23	31.00
35	26.63	26.63	26.49	26.67	26.66	26.53
40	23.32	23.32	23.23	23.44	23.43	23.34
45	20.73	20.73	20.67	20.78	20.78	20.72

Table 8.2

	${}_n\gamma_l$ values					
	HB 1			UTD 124A′		
n	$l=2$	$l=4$	$l=18$	$l=2$	$l=4$	$l=18$
1	477.0	483.3	320.7	483.3	489.9	318.3
5	465.1	465.1	475.0	478.8	477.8	466.1
10	462.2	462.5	462.5	443.8	443.9	451.9
15	469.0	468.9	469.0	487.4	486.6	476.1
20	467.1	466.6	465.7	449.4	449.8	451.5
25	467.3	468.6	468.4	495.0	496.6	489.5
30	463.7	463.5	463.3	454.1	453.8	453.2
35	465.9	462.7	465.5	499.6	499.3	499.6
40	465.9	465.9	463.2	457.2	460.9	458.2
45	472.2	472.2	470.2	497.8	497.8	495.6

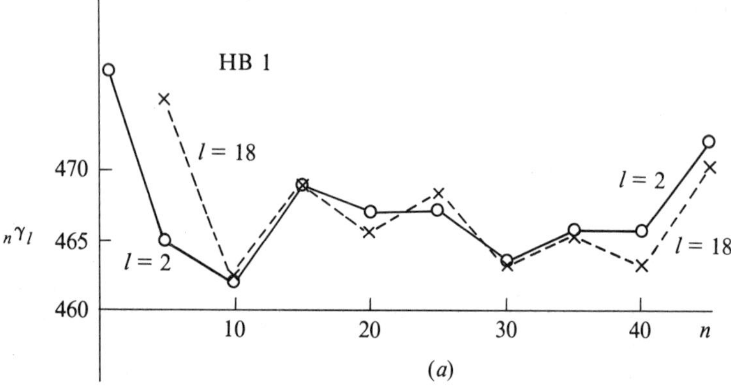

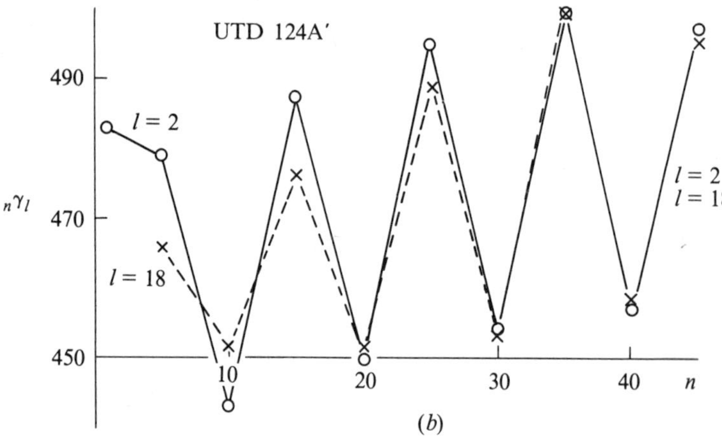

Fig. 8.2. (*a*) Graphs of $_n\gamma_l$ against n for $l = 2, 18$ for HB 1. (*b*) Graphs of $_n\gamma_l$ against n for $l = 2, 18$ for UTD 124A′.

$M(r) = r^2\rho^{1/2}c_S^{1/2}$ is not small, and $d\mu/dr$, $d\rho/dr$ are not large, throughout the range of r. We have next to consider the distribution of high overtone frequencies when μ or ρ or one of their derivatives contains a discontinuity in the range of r. In Fig. 8.3 (Lapwood & Sato, 1979) we exhibit the pattern of variation of $S = (\gamma_n\omega_2/\pi) - n$ with n for PEM–A (Parametrically simple Earth-model A) of Dziewonski, Hales & Lapwood (1975). This model has six internal surfaces of discontinuity. We want to be able to explain such patterns.

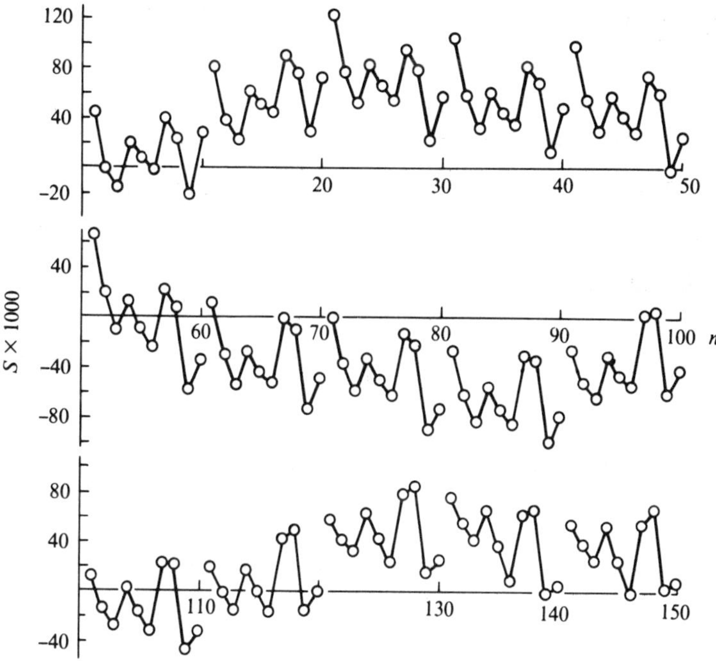

Fig. 8.3. Graph of $S = (\gamma\omega/\pi) - n$ against n for $n = 1, 2, 3, \ldots, 150$, $(l = 2)$.

8.3 The distribution of high frequencies of toroidal overtones for a sphere with internal discontinuities

When there are discontinuities of material constants ρ and μ within the elastic sphere, the equations (8.1.22) must be replaced by separate sets for the different layers between discontinuities, and correct interface conditions used to link the solutions in different layers. To simplify our working we will suppose that there is one surface of discontinuity only, at $r = c$, $b < c < a$. Let the suffix 1 refer to the region $b < r < c$, and the suffix 2 to the region $c < r < a$. Then after the Liouville transformation we get, instead of (8.1.24) and (8.1.25),

$$Z_1'' + Q_1 Z_1 = 0,\ Q_1 = \omega^2 - q_1(s),\ q_1(s) = \left\{\frac{M_1''}{M_1} + \frac{(l+2)(l-1)c_S^2(s)}{r^2(s)}\right\}_1, \tag{8.3.1}$$

$$Z_2'' + Q_2 Z_2 = 0,\; Q_2 = \omega^2 - q_2(s),\; q_2(s) = \left\{\frac{M_2''}{M_2} + \frac{(l+2)(l-1)c_S^2(s)}{r^2(s)}\right\}_2. \tag{8.3.2}$$

At the inner boundary ($r = b$), $s = 0$, and

$$Z_1' - h_1 Z_1 = 0 \qquad \text{at } s = 0, \qquad \text{where } h_1 = (M_1'/M_1)_{r=b}; \tag{8.3.3}$$

at the surface ($r = a$)

$$s = \gamma = \int_b^a \mathrm{d}r/c_S(r), \tag{8.3.4}$$

and

$$Z_2' - h_2 Z_2 = 0 \qquad \text{at } s = \gamma, \qquad \text{where } h_2 = (M_2'/M_2)_{r=a}. \tag{8.3.5}$$

At the interface ($r = c$), W and $\mu(\dot{W} - W/r)$ must be continuous. (8.3.6)

Let $s = s_1 = \int_b^a \mathrm{d}r/c_S$ at the interface. The conditions (8.3.6) become

$$\left.\begin{aligned} &\frac{Z_1}{(\rho_1 c_{S1})^{1/2}} = \frac{Z_2}{(\rho_2 c_{S2})^{1/2}}, && (8.3.7)\\ &\frac{\mu_1}{(\rho_1 c_{S1})^{1/2}}\left\{\frac{Z_1'}{c_{S1}} - \frac{Z_1 M_1'}{c_{S1} M_1}\right\} = \frac{\mu_2}{(\rho_2 c_{S2})^{1/2}}\left\{\frac{Z_2'}{c_{S2}} - \frac{z_2 M_2'}{c_{S2} M_2}\right\}, && (8.3.8)\end{aligned}\right\} \text{at } s = s_1.$$

We now introduce the modified *Prüfer substitution*

$$\cot\phi = Q^{-1/2} Z'/Z. \tag{8.3.9}$$

For discussion of the distribution of eigenvalues of this system we need consider only ϕ_1 and ϕ_2. To satisfy (8.3.3) we take $\phi_1(s, \omega)$ such that

$$\cot\phi_1(0, \omega) = h_1 Q_1^{-1/2}(0), \tag{8.3.10}$$

and to satisfy (8.3.5) we take $\phi_2(s, \omega)$ such that

$$\cot \phi_2(\gamma, \omega) = h_2 Q_2^{-1/2}(\gamma). \tag{8.3.11}$$

From (8.3.1) and (8.3.2) we obtain the differential equations for ϕ_1 and ϕ_2:

$$\left.\begin{aligned} \phi_1' &= (\omega^2 - q_1)^{1/2} - \tfrac{1}{4} q_1' \sin 2\phi_1 (\omega^2 - q_1)^{-1}, \\ \phi_2' &= (\omega^2 - q_2)^{1/2} - \tfrac{1}{4} q_2' \sin 2\phi_2 (\omega^2 - q_2)^{-1}. \end{aligned}\right\} \tag{8.3.12}$$

For large ω, these can be recast in the form:

$$\left.\begin{aligned} \phi_1' &= \omega - \tfrac{1}{2} q_1(s)/\omega + O(1/\omega^2), \\ \phi_2' &= \omega - \tfrac{1}{2} q_2(s)/\omega + O(1/\omega^2). \end{aligned}\right\} \tag{8.3.13}$$

Integrating, we obtain

$$\phi_1(s_1, \omega) \approx \phi_1(0, \omega) + \omega s_1 - (1/2\omega) \int_0^{s_1} q_1(s)\,\mathrm{d}s, \tag{8.3.14}$$

$$\phi_2(\gamma, \omega) \approx \phi_2(s_1, \omega) + \omega(\gamma - s_1) - (1/2\omega) \int_{s_1}^{\gamma} q_2(s)\,\mathrm{d}s. \tag{8.3.15}$$

We seek an approximation valid for high frequencies, so we assume $\omega^2 \gg q_1(s)$ and $\omega^2 \gg q_2(s)$ throughout the range of s. Note that this implies that l is not too large. Then from (8.3.1), Z_1 will behave like $\cos(\omega s + \delta)$, and Z_1' will be of the order ωZ. In a zero-order approximation the boundary condition (8.3.3) is $Z_1' = 0$ at $s = 0$, that is, $\cos \phi_1(0) = 0$.

We select the phase of ϕ_1 to be $\frac{1}{2}\pi$ at $s = 0$. From (8.3.10), the first-order approximation is $\phi_1(0, \omega_n) = \frac{1}{2}\pi - h_1/\omega_n$, where ω_n is the frequency of the nth mode. We define that mode as the one in which the phase at $s = \gamma$ is in advance by $n\pi$ of the phase at $s = 0$. Then by a similar argument

$$\phi_2(\gamma, \omega_n) = n\pi + \tfrac{1}{2}\pi - h_2/\omega_n, \tag{8.3.16}$$

approximately.

We have still to satisfy the conditions at the interface. Using $Z'/Z = Q^{1/2} \cot \phi$ in (8.3.8), taking the ratio of Z_1/Z_2 at the interface from (8.3.7) and neglecting terms of order $1/\omega$, we get

$$(\mu_1/c_{S1}) Q_{1c}^{1/2} \cot \phi_{1c} = (\mu_2/c_{S2}) Q_{2c}^{1/2} \cot \phi_{2c}. \tag{8.3.17}$$

We have assumed that ϕ_{1c} and ϕ_{2c} (where the suffix c indicates the value at the interface) are not such that both $\cot\phi_{1c}$ and $\cot\phi_{2c}$ are of order $1/\omega_n$. (8.3.17) may be re-stated as

$$\rho_1 c_{S1} \sin\alpha_{1c} \cot\phi_{1c} = \rho_2 c_{S2} \sin\alpha_{2c} \cot\phi_{2c}, \tag{8.3.18}$$

where we have put

$$Q_{1c} = \omega^2 \sin^2\alpha_{1c}, \qquad Q_{2c} = \omega^2 \sin^2\alpha_{2c}. \tag{8.3.19}$$

If we now introduce into (8.3.18) the reflection coefficient (see § 8.3.2)

$$R' = \frac{\rho_2 c_{S2} \sin\alpha_{2c} - \rho_1 c_{S1} \sin\alpha_{1c}}{\rho_2 c_{S2} \sin\alpha_{2c} + \rho_1 c_{S1} \sin\alpha_{1c}}, \tag{8.3.20}$$

and

$$\chi_1 = \phi_{1c} - \tfrac{1}{2}\pi, \qquad \chi_2 = n\pi + \tfrac{1}{2}\pi - \phi_{2c}, \tag{8.3.21}$$

we find

$$\sin(\chi_2 + \chi_1) + R' \sin(\chi_2 - \chi_1) = 0, \tag{8.3.22}$$

where from (8.3.14), (8.3.15) and (8.3.16)

$$\left.\begin{aligned} \chi_1 &= \omega_n s_1 - (1/2\omega_n)\int_0^{s_1} q_1(s)\,ds + O(1/\omega),\\ \chi_2 &= \omega_n(\gamma - s_1) - (1/2\omega_n)\int_{s_1}^{\gamma} q_2(s)\,ds + O(1/\omega). \end{aligned}\right\} \tag{8.3.23}$$

If we make R' zero, (8.3.22) reduces to

$$\chi_1 + \chi_2 = n\pi, \tag{8.3.23$'$}$$

which is the frequency equation when we allow for Earth curvature or radial inhomogeneity, but neglect reflections at interfaces.

8.3.1 *Flat layering approximation*

If as a first approximation we neglect q_1 and q_2, then

$$\left.\begin{aligned} \chi_1 &= \omega_n s_1 = \omega_n \theta, &\quad \text{where } \theta &= \int_b^c dr/c_S,\\ \chi_2 &= \omega_n(\gamma - s_1) = \omega_n \phi, &\quad \text{where } \phi &= \int_c^a dr/c_S = \gamma - \theta, \end{aligned}\right\} \tag{8.3.24}$$

$$R' = \frac{\rho_2 c_{S2} - \rho_1 c_{S1}}{\rho_2 c_{S2} + \rho_1 c_{S1}} = R, \tag{8.3.25}$$

the coefficient of reflection at *normal* incidence from region 2 onto the interface; (8.3.22) becomes, for general ω,

$$\sin \omega(\phi + \theta) + R \sin \omega(\phi - \theta) = 0. \tag{8.3.26}$$

This is the formula given by Anderssen (1977): the roots ω_n of this equation are the eigenfrequencies. If R is small, then for large n

$$\omega_n \simeq \frac{1}{\phi + \theta} n\pi + (-1)^{n+1} \frac{R}{\phi + \theta} \sin \frac{\phi - \theta}{\phi + \theta} n\pi. \tag{8.3.27}$$

If $R = 0$, we recover $\omega_n \simeq n\pi/\gamma$, the asymptotic relation for a shell without internal reflections. Thus the second term on the right of (8.3.27) expresses the effect of internal relections: ω_n are no longer equally spaced, but show an irregular pattern continually repeated. The name '*solotone*' meaning 'single frequency', was given to this effect (McNabb *et al.* 1976), measured approximately by $(\gamma\omega_n/\pi) - n = S$, where

$$S = (-1)^n \frac{R}{\pi} \sin \frac{\theta - \phi}{\theta + \phi} n\pi. \tag{8.3.28}$$

Since (8.3.27) does not contain l, our approximation, in neglecting q_1, q_2, has wiped out the effect of the Earth's curvature.

8.3.2 *Allowance for Earth curvature*

We now seek a better approximation to the solution of the system (8.3.1) to (8.3.5), but to simplify working and interpretation we will take the regions 1 and 2 to be uniform. Then for each shell

$$\left.\begin{aligned} M &= r^2 \rho^{1/2} c_S^{1/2}, \qquad & M' &= \mathrm{d}M/\mathrm{d}s = c_S \mathrm{d}M/\mathrm{d}r \\ & & &= 2r\rho^{1/2} c_S^{3/2}, \\ & & M'' &= 2\rho^{1/2} c_S^{5/2}, \\ M'/M &= 2c_S/r, & M''/M &= 2c_S^2/r^2, \end{aligned}\right\} \tag{8.3.29}$$

$$\begin{gathered} h_1 = 2c_S/b, \qquad h_2 = 2c_S/a, \\ q = \{2 + (l+2)(l-1)\} c_S^2/r^2 = l(l+1) c_S^2/r^2. \end{gathered} \tag{8.3.30}$$

But if, corresponding to (8.3.19), we put $Q = \omega^2 \sin^2 \alpha$, then

$$\omega^2 \sin^2 \alpha = \omega^2 - l(l+1)c_S^2/r^2,$$

and

$$\cos \alpha = Lc_S/\omega r, \qquad \text{where } L^2 = l(l+1). \tag{8.3.31}$$

Thus

$$dr = (Lc_S/\omega) \sec \alpha \tan \alpha \, d\alpha, \tag{8.3.32}$$

and from (8.3.12), neglecting the second term on the right,

$$\begin{aligned} \chi_1 &= \phi_1(s_1, \omega) - \phi_1(0, \omega) \\ &\approx \int_0^{s_1} (\omega^2 - q_1)^{1/2} \, ds \\ &= L \int_{\alpha_1(b)}^{\alpha_1(c)} \tan^2 \alpha_1 \, d\alpha_1 \\ &= L(\tan \alpha_1 - \alpha_1)_b^c. \end{aligned} \tag{8.3.33}$$

This is the phase change in region 1. Similarly,

$$\phi_2(\gamma, \omega) - \phi_2(s_1, \omega) \approx L(\tan \alpha_2 - \alpha_2)_c^a \tag{8.3.34}$$

is the phase change in region 2.

We now construct an interpretation of these formulae, (8.3.33) and (8.3.34). According to geometrical ray theory, any ray may be specified by a *ray parameter* p, which has the following properties: at each point of the ray $p = r \sin i/v$, where i is the angle between the ray direction and the radius to the point and v the local wave-speed (Fig. 8.4(a)); at the surface $p = a \sin i_0/v_0 = dT/d\Delta$, where T is the time of arrival at epicentral distance $a\Delta$; at the deepest point of the ray $i = \frac{1}{2}\pi$ and $r_d = v_d p$. (These formulae are correct whether the ray consists of smooth or broken curves.)

Now consider, for high frequency ω and high degree l, an equivalence between a normal mode and a ray with the same wave-length:

$$p = dT/d\Delta = a/V_s, \tag{8.3.35}$$

where V_s is the speed along the surface. If ω is the angular frequency and λ_s the wave-length, $\lambda_s = V_s 2\pi/\omega$. Thus the number of waves in a

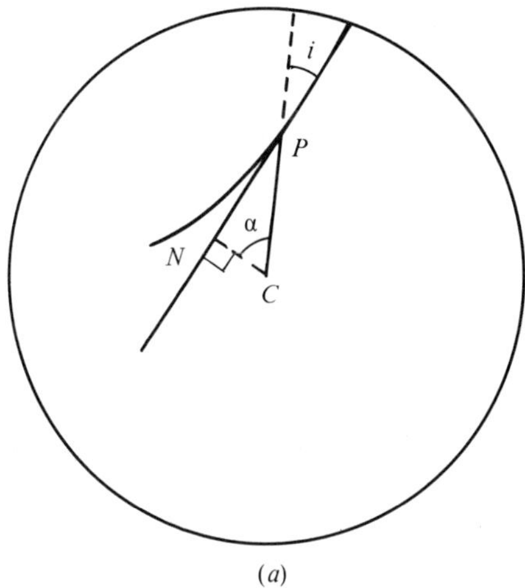

(a)

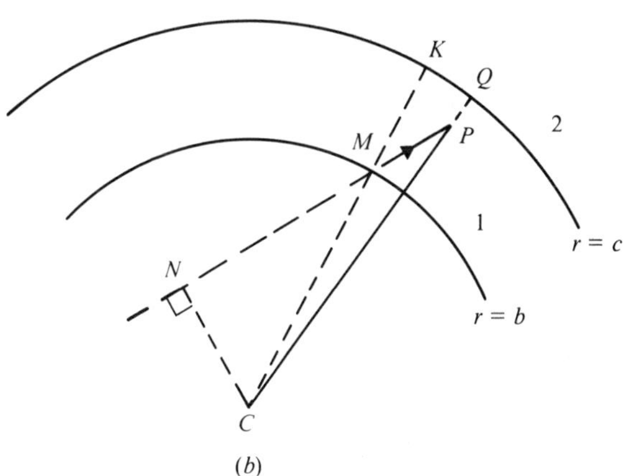

(b)

Fig. 8.4. (a) Ray in sphere. (b) Ray in layered sphere.

complete meridian is

$$2\pi a/(V_s 2\pi/\omega) = a\omega/V_s = \omega p.$$

The mode of degree l has l oscillations in the complete meridian, but they are not of equal length. Except in the neighbourhood of the poles they are asymptotically, for large l, of equal length, but those in the neighbourhood of each pole have a total additional length of a

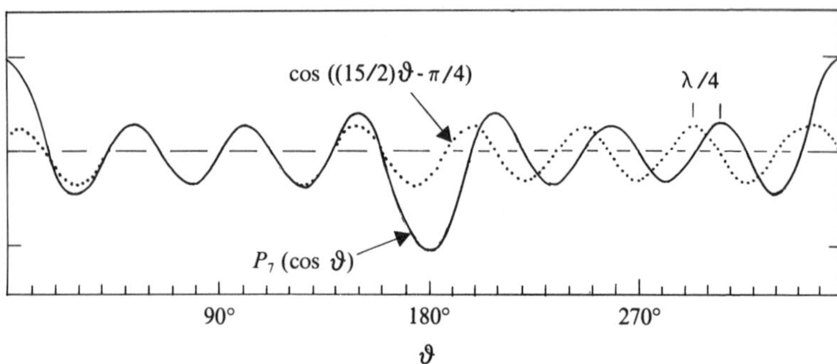

Fig. 8.5. A comparison of the phase of P_7 with that of its asymptotic representation. (From Brune *et al.*, 1961.)

quarter wave-length (see Fig. 8.5 due to Brune, Nafe & Alsop (1961)). Thus in terms of the wave-length as measured everywhere except near the poles, there are $l+\frac{1}{2}$ waves along the meridian, and so

$$\omega p = l + \tfrac{1}{2}. \tag{8.3.36}$$

This can also be shown by asymptotic expansion of $P_l(\cos\vartheta)$ (Ben–Menahem, 1964); see also the elegant paper of Gilbert (1976).

We next turn back to (8.3.31), and note that, for large l, $l(l+1) \simeq (l+\frac{1}{2})^2$, so that $\cos\alpha \simeq vp/r$ and $\alpha \simeq \frac{1}{2}\pi - i$ ($\angle\, NCP$ in Fig. 8.4(a)).

Now consider the formula (8.3.33) for χ_1. For any P in region 1 (Fig. 8.4(b)),

$$L(\tan\alpha_1 - \alpha_1)_b^r = \omega p \tan\alpha_1\,\big|_M^P - 2\pi L(\tfrac{1}{2}\alpha_1/\pi)_K^Q. \tag{8.3.37}$$

Here the first term is the change of phase along MP, while the second is the corresponding change of phase in the l-mode. Thus $L(\tan\alpha_1 - \alpha_1)_b^r$ is the phase gain of body wave over mode between M and P. So $\chi_1 + \chi_2$ is the total phase gain in one oblique transit of the shell by the S wave. When it is placed equal to $n\pi$ we have constructive interference after *double* transit.

The formula (8.3.20) for R' now enables us to identify R' as the coefficient of reflection when a ray from region 2 strikes the interface obliquely.

Finally, the frequency equation (8.3.22) gives the pattern of eigenfrequencies when we allow both for internal reflections (at a spherical surface of discontinuity of material constants) and also for Earth-curvature.

8.3.3 *Comparison of approximations*

If we select a problem in which an exact solution is available, so that eigenfrequencies can be calculated to any desired precision, we can compare the different approximations. Such a problem is that of a spherical shell composed of any number of uniform layers: the frequency equation can be obtained in terms of spherical Bessel functions (Sato & Lapwood, 1977a, b).

For computational convenience, since we can obtain ω_n with high precision, we will invert the formula $\omega_n \simeq n\pi/\gamma$, which is obtained by neglecting internal reflections and Earth-curvature, and define

$$n_1 = \gamma\omega_n/\pi. \qquad (8.3.38)$$

Then $n - n_1$ is a measure of the accuracy of the simple formula (8.3.38).

Secondly, we refer to the approximate formula (8.3.23′), and define

$$n_2 = (\chi_1 + \chi_2)/\pi. \qquad (8.3.39)$$

$\chi_1 + \chi_2$ has the following physical interpretation.
From Fig. 8.6,

$$\begin{aligned} 2\chi_1 &= 2L(\tan\alpha_1 - \alpha_1)_b^c \\ &= \omega(2FG/v_{S1} - 2p\alpha_1|_G^F). \end{aligned}$$

Similarly

$$2\chi_2 = \omega(2EF/v_{S2} - 2p\alpha_2|_F^E).$$

Thus

$$2(\chi_1 + \chi_2) = \omega(T - \Delta\mathrm{d}T/\mathrm{d}\Delta), \qquad (8.3.40)$$

where Δ is epicentral distance EK, and T travel-time. We may measure the accuracy of formula (8.3.39) by $n_2 - n$.

The flat layering approximation for the solotone effect is given by S in (8.3.28), viz.

$$S = (-1)^n \frac{R}{\pi} \sin \frac{\theta - \phi}{\theta + \phi} n\pi. \qquad [(8.3.28)]$$

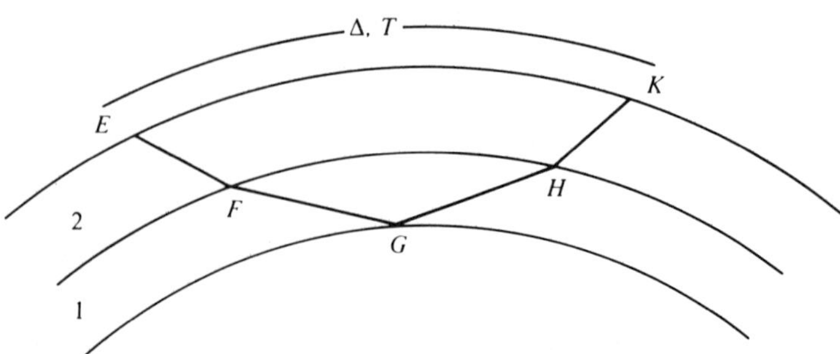

Fig. 8.6. Interpretation of $\chi_1 + \chi_2$.

For the corresponding approximation which takes account of Earth-curvature, we replace (8.3.39) by

$$\chi_1 + \chi_2 = n\pi + \delta_n \tag{8.3.41}$$

in (8.3.22), and obtain, after using $\chi_1 + \chi_2 \approx n\pi$ in the small term,

$$\delta_n \approx \frac{R' \sin 2\chi_1}{1 + R' \cos 2\chi_1}. \tag{8.3.42}$$

We now consider a very simple Earth-model, defined as follows:

Inner radius:	$r = b = 3\,470$ km.
Surface of discontinuity:	$r = c = 5\,960$ km.
Outer radius:	$r = a = 6\,370$ km.

In region

(1) $b < r < c$, $\quad c_{S1} = 6.58$ km/s, $\quad \rho_1 = 5.35$ g/cm^3,
(2) $c < r < a$, $\quad c_{S2} = 4.46$ km/s, $\quad \rho_2 = 3.77$ g/cm^3.

Thus the coefficient of reflection for a ray going from (2) to impinge normally on the surface of discontinuity is

$$R = \frac{\rho_2 c_{S2} - \rho_1 c_{S1}}{\rho_2 c_{S2} + \rho_1 c_{S1}} = -0.354.$$

This model has a single internal surface of discontinuity at depth 410 km.

In Fig. 8.7 we show $n_1 - n$ and $n_2 - n$ graphed against n, for $l = 2$ in (a) and for $l = 15$ in (b). Neither approximation allows for the

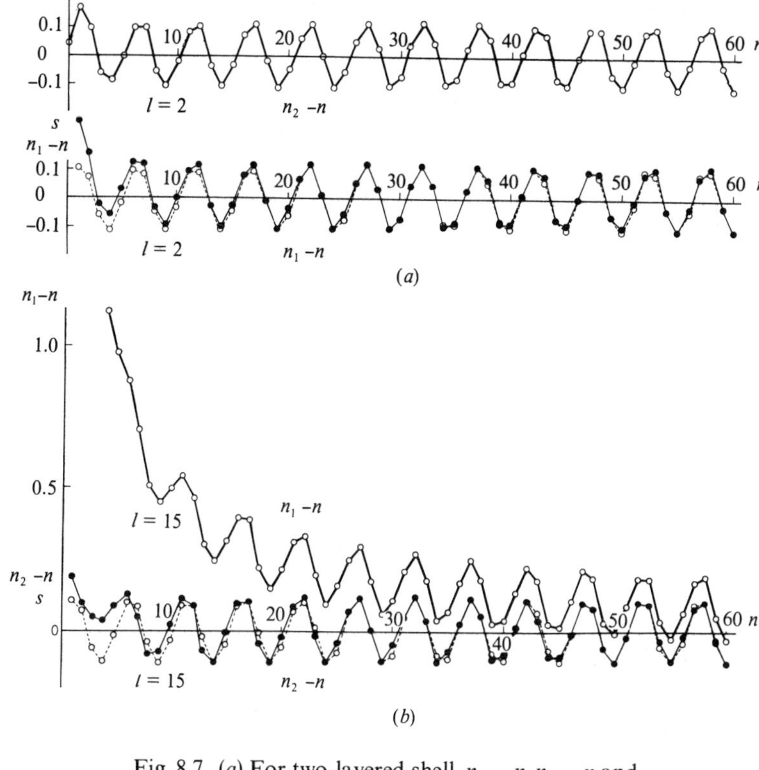

Fig. 8.7. (*a*) For two-layered shell, $n_2 - n, n_1 - n$ and $S = (-)^n (R/\pi) \sin\{(\theta - \phi)n\pi/(\theta + \phi)\}$ graphed against n for $l = 2$ and $n = 0, 1, 2, \ldots, 60$.
—○— joins values of $n_2 - n$,
—●— joins values of $n_1 - n$,
- -○- - joins values of S.
(*b*) For two-layer shell, $n_1 - n, n_2 - n$ and $S = (-)^n (R/\pi) \sin\{(\theta - \phi)n\pi/(\theta + \phi)\}$ graphed against n for $l = 15$ and $n = 0, 1, 2, \ldots, 60$.
—○— joins values of $n_1 - n$,
—●— joins values of $n_2 - n$,
- -○- - joins values of S.

solotone effect; but that is graphed in (*a*) and (*b*) by S (given in (8.3.28)), and it can be seen that S very nearly accounts for the whole error $n_2 - n$. $n_1 - n$ incorporates an additional error, especially for small n and larger l, due to failure to allow for Earth-curvature.

8.4 Patterns of eigenfrequencies

8.4.1 *The pattern of eigenfrequencies for a two-layer shell*

In Fig. 8.7(*a*) the points corresponding to adjacent frequencies were joined by broken lines. In Fig. 8.8 we join up the same set of points in a new way and discover a remarkable and significant pattern. In this figure we draw five continuous curves through the sets of points corresponding to the following values of n:

$$(3, 8, 13, \ldots, 58), \quad (4, 9, 14, \ldots, 59), \quad (5, 10, 15, \ldots, 60),$$
$$(6, 11, 16, \ldots, 56), \quad (7, 12, 17, \ldots, 57).$$

The points of one set lie on a smooth curve which changes slowly in direction. This can be understood by reference to Anderssen's formula (8.3.28) for S, which is nearly equal to $n_2 - n$. Here $(\theta - \phi)/(\theta + \phi) = \frac{3}{5} + \varepsilon$, where

$$\varepsilon = 0.0091, \tag{8.4.1}$$

which is small. If we now write $n = 5p + q$, where $p = 0, 1, 2, \ldots$, and $q = 0, 1, 2, 3, 4$, then

$$S = (-1)^q (R/\pi) \sin(5\varepsilon p\pi + \delta), \qquad \text{where } \delta = (\tfrac{3}{5} + \varepsilon) q\pi. \tag{8.4.2}$$

The five curves of Fig. 8.8 are given by (8.4.2) with $q = 0, 1, 2, 3, 4$. The amplitudes R/π and periods $2/(5\varepsilon)$ in p are the same for all curves: the phase is given by (8.4.2).

Fig. 8.8 is drawn for $l = 2$. For $l = 5, 10, 15$ the same pattern is found, with small variations. The pattern is not sensitive to the Legendre degree l. The amplitude of the solotone oscillation depends on R, which is determined by the contrast in impedance at the surface of discontinuity. But the *period* depends on the ratio of $\theta - \phi$ to $\theta + \phi$. If this ratio can be expressed as $\pm i/j + \varepsilon$, where ε is small and i, j are small integers such that $i + j$ is even, there will be j

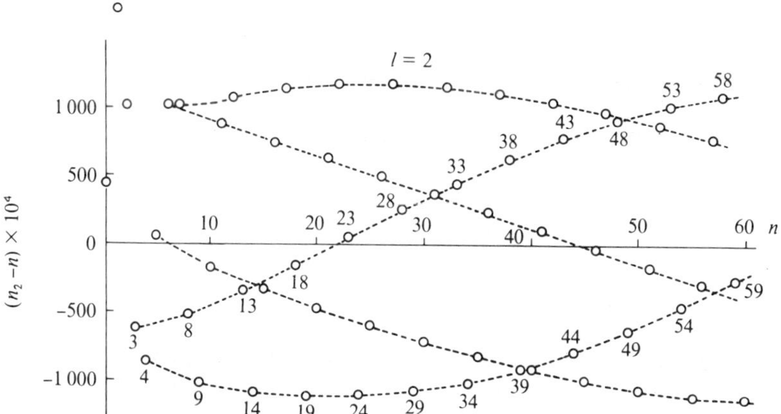

Fig. 8.8. Pattern of eigenfrequencies for two-layer shell for $l = 2, n = 1, 2, 3, \ldots, 60$. The points are the same as in Fig. 8.7(*a*), but are here shown to lie on smooth sinusoidal lines.

sine curves on which the points lie in the diagram corresponding to Fig. 8.8.

8.4.2 *The shift of frequency pattern when layer thicknesses change*

In the two-layer sphere studied above the period of the solotone depended on $(\theta - \phi)/(\theta + \phi)$. This suggests that a small change in the travel-times of shear waves through the different layers may make a comparatively large change in the pattern of eigenfrequencies. Lapwood & Sato (1978) tested this conjecture by a numerical experiment in which an averaged PEM–A Earth-model, with two internal discontinuities, at approximately 400 km and 600 km depth, was perturbed by moving the interfaces. The standard model A (averaged PEM–A) was specified as follows:

$$R_1 = \frac{\rho_2 c_{S2} - \rho_1 c_{S1}}{\rho_2 c_{S2} + \rho_1 c_{S1}}, \qquad R_2 = \frac{\rho_3 c_{S3} - \rho_2 c_{S2}}{\rho_3 c_{S3} + \rho_2 c_{S2}};$$

$r_3 = 6368$ km $\quad c_{S3} = 4.51$ km/s $\quad \rho_3 = 3.42$ g/cm^3 $\quad R_2 = -0.1516$

$r_2 = 5951$ km $\quad c_{S2} = 5.30$ km/s $\quad \rho_2 = 3.95$ g/cm^3 $\quad R_1 = -0.2367$

$r_1 = 5701$ km $\quad c_{S1} = 6.77$ km/s $\quad \rho_1 = 5.01$ g/cm^3

$r_0 = 3485.7$ km

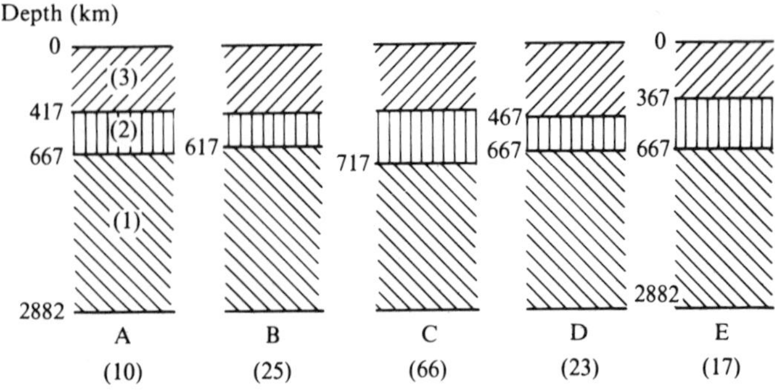

Fig. 8.9. Scheme of layer thicknesses in models A, B, C, D, E; A being an averaged PEM–A. Depths of discontinuities are shown on the left side of each section. Figures in brackets below the letters A to E show the recurrence periods (in n) of the patterns of eigenfrequencies for the five models.

Perturbations from model A were arranged as follows:

In B the lower discontinuity is raised by 50 km.
In C the lower discontinuity is lowered by 50 km.
In D the upper discontinuity is lowered by 50 km.
In E the upper discontinuity is raised by 50 km.

These five arrangements of layers are shown in Fig. 8.9. The corresponding patterns of eigenfrequencies are shown in Figs. 8.10(a–e) for $n = 1, 2, 3, \ldots, 60$. We notice remarkable changes in the recurrence period in n. This period is 10 for A, 25 for B, 23 for D and 17 for E. At first inspection there seems to be no recurrence in C; but closer inspection shows that values do repeat – though with opposite sign, when n is increased by 33. Thus a recurrence period of 66 in n exists, and would be seen if a longer series of values of n was used.

Here, for three layers, we use the form to which (A.5.29) reduces when $m = 3$, viz.

$$S \simeq (R_1/\pi) \sin 2\chi_1 - (R_2/\pi) \sin 2\chi_3 . \tag{8.4.3}$$

(See also Lapwood & Sato (1977)). If we write $2\chi_1 = n\alpha_1$, $2\chi_3 = n\alpha_2$, defining α_1 and α_2, then there will be recurrence with period j in n if there is an integer j such that both $j\alpha_1$ and $j\alpha_2$ are nearly integral multiples of 2π. We find the following results:

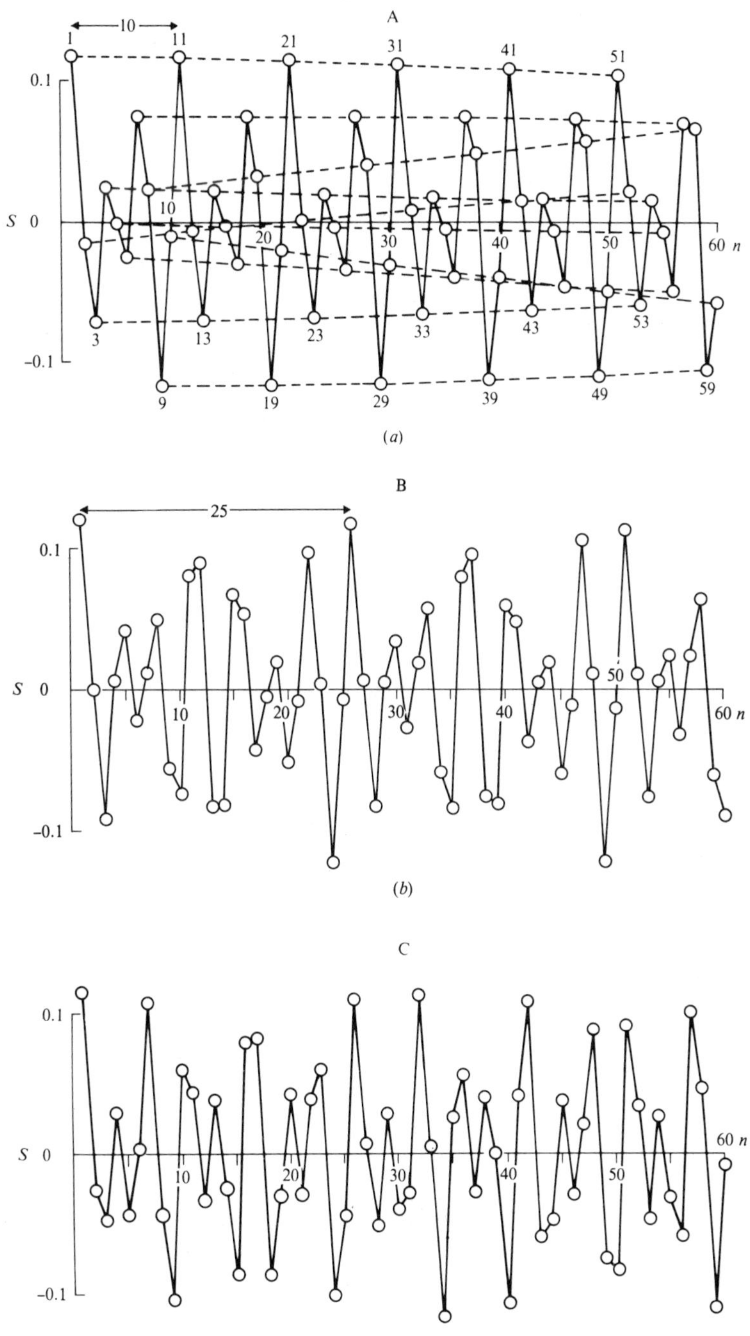

(Cont.)

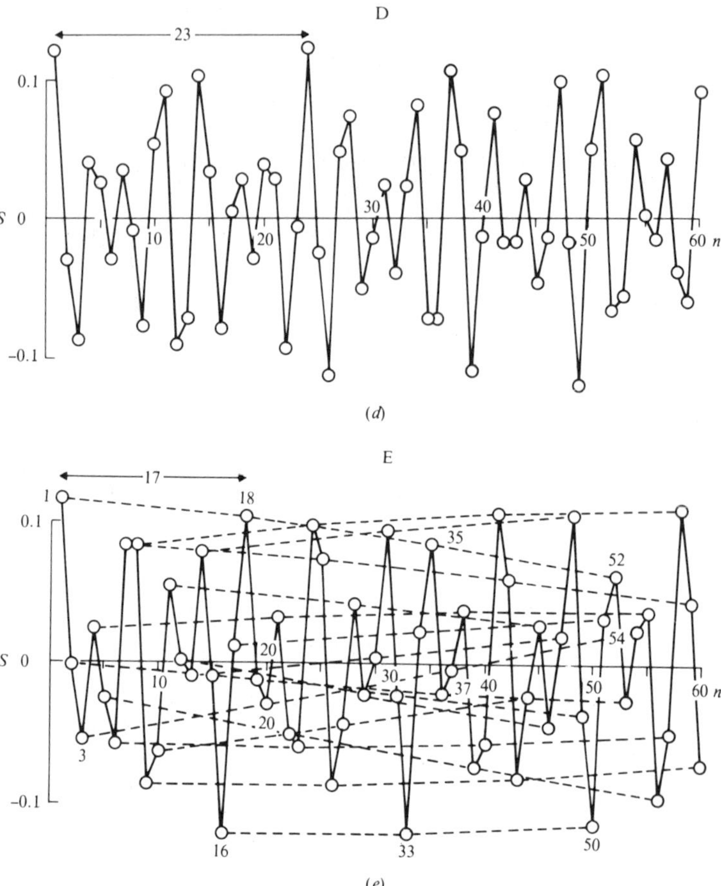

Fig. 8.10. (*a*) Graph of S against n for model A; the recurrence period of 10 in n shows clearly. The ten almost flat lineations in the pattern of eigenfrequencies are shown by dashed lines. That beginning at $n = 1$ carries the numbers $(1, 11, 21, 31, 41, 51)$ of the modes whose frequencies lie on it. Similarly for the lines starting at $n = 3$ and $n = 9$ (to avoid confusion the other dashed lines carry no numbers). (*b*) Graph of S against n for model B; the recurrence period is 25. (*c*) Graph of S against n for model C. After an interval of 33 in n, the pattern of eigenfrequencies repeats with reversed sign; this can be seen in the graph. The recurrence period of 66 is too long to show in the graph. (*d*) Graph of S against n for model D; the recurrence period is 23. (*e*) Graph of S against n for model E; the recurrence period is 17. The 17 dashed lines mark lineations; they are not as flat as those in (*a*) (model A), for reasons given in the text. The dashed line starting at $n = 1$ carries the numbers of those modes whose frequencies lie on it $(1, 18, 35, 52)$. Similarly for the lineations starting at $n = 3$ and $n = 16$.

Model	j	$j\alpha_1/\pi$	$j\alpha_2/\pi$
A	10	14.018	3.961
B	25	35.994	9.946
C	66	90.037	26.029
D	23	32.128	10.167
E	17	23.916	5.947

The alignments corresponding to these values of j are indicated in Figs. 8.10(*a*) and 8.10(*e*), but not in the other figures, since the addition of large numbers of roughly parallel line segments would make the figures confusing. We may note that the fact that $j\alpha_1/\pi$ and $j\alpha_2/\pi$ diverge more from even integers for E than for A means that the alignments in E diverge further from parallels to the n-axis than those in A.

8.5 The pattern of frequencies of radial overtones for the realistic Earth-model PEM–A

From simple Earth-models, we now turn to consider a realistic Earth-model, PEM–A (parametrically simple Earth-model A) (Dziewonski, Hales & Lapwood, 1975). This is defined as follows:

Layer	Interface (radius, in km)	Reflection coefficient	Radial travel-time within layer (in s)	
	6 368.0			
Crust 7			3.098 59 (7)	(8.5.1)
	6 357.0	$R_6 = -0.044\,91$		
Crust 6			1.333 33 (6)	
	6 352.0	$R_5 = -0.172\,04$		
5			13.107 00 (5)	
	6 291.0	$R_4 = 0.034\,84$		
4			32.253 61 (4)	
	6 151.0	$R_3 = -0.013\,60$		
3			42.770 49 (3)	
	5 951.0	$R_2 = -0.054\,25$		
2			47.174 23 (2)	
	5 701.0	$R_1 = -0.090\,85$		
Lower mantle 1			327.800 71 (1)	
	3 485.7			
			467.537 96 (Total)	

(Note that the numbering of layers and reflection coefficients proceeds *upwards* from the core boundary.) In (8.5.1) R_j is calculated from

$$\frac{\rho'_{j+1} v'_{S,j+1} - \rho_j v_{S,j}}{\rho'_{j+1} v'_{S,j+1} + \rho_j v_{S,j}}, \tag{8.5.2}$$

where ρ_j is density and $v_{S,j}$ shear velocity adjacent to the $(j, j+1)$ interface in the jth layer, and $\rho'_{j+1}, v'_{S,j+1}$ adjacent to the same interface in the $(j+1)$th layer.

As shown in Appendix § A.5, the solotone correction for a shell of m layers is

$$S \simeq -\frac{1}{\pi} \sum_{j=1}^{m-1} R_j \sin\left(\frac{2n\pi}{\gamma} \sum_{p=j+1}^{m} \chi'_p\right), \tag{8.5.3}$$

where

$$\chi'_p = (T - \Delta \mathrm{d}T/\mathrm{d}\Delta)_{p\text{th layer}} \tag{8.5.4}$$

and

$$\gamma = \sum_{p=1}^{m} \chi'_p. \tag{8.5.5}$$

S has been computed for $l = 2$ and $n = 1, 2, \ldots, 150$. The values thus found are graphed against n in Fig. 8.3. We observe immediately that there is a pattern of period 10 in n, which changes slowly as n proceeds from 1 to 150. We have taken this large range of n to bring out another period of S, which is shown in the long sinusoidal arc of period about 110 in n. We now look for explanations of these two recurrence periods.

Substituting numerical values from (8.5.1) into (8.5.3) and retaining only such precision as is adequate to our purpose, we find

$$\begin{aligned} S = (1/\pi)\{&0.091 \sin 2n\pi(0.2989) + 0.054 \sin 2n\pi(0.1980) \\ &+ 0.014 \sin 2n\pi(0.1065) - 0.035 \sin 2n\pi(0.0375) \\ &+ 0.172 \sin 2n\pi(0.0095) + 0.045 \sin 2n\pi(0.0066)\}. \end{aligned} \tag{8.5.6}$$

The coefficients of the sines are $R_1, R_2, \ldots, R_6$, respectively. The largest amplitudes are those of the fifth and first terms, while the contribution of the third term is very small.

If we now ask what lowest values of n will produce almost exact integral multiples of 2π in the successive arguments of the sines, we find, respectively

$$10, 10, 10, 80, 105, 150. \tag{8.5.7}$$

Thus the first three terms of (8.5.6) have the same sign and almost the same period in n, and together account for the dominant period of 10 in n in the pattern of eigenfrequencies shown in Fig. 8.3. The first term contains R_1, which is the coefficient of reflection at the 667 km discontinuity; the second contains R_2, which is the coefficient of reflection at the 417 km discontinuity, while the third contains R_3, corresponding to a small discontinuity at 217 km.

The fifth term has R_5 as factor, corresponding to reflection at the base of the crust. Thus the shape of the pattern of eigenfrequencies seems to be determined mainly – in fact almost completely – by resonances among reflections at the 667 km and 417 km discontinuities and the Mohorovičić discontinuity.

8.6 Solotone effect due to internal resonances

We now seek a physical explanation of the solotone effect. It is already clear that it must involve a beating of almost equal frequencies.

8.6.1 *Resonance relations*

Our formula for S for a two-layer shell (8.3.28) may be rewritten

$$S = (-1)^n (R/\pi) \sin\{n\pi(T_1 - T_2)/T\}, \tag{8.6.1}$$

where T_1 is the time taken for a ray to cross layer 1, and T_2 to cross layer 2, and $T = T_1 + T_2$. Then if $T_1 - T_2$ divides into T almost an integral number of times, say k, the pattern of frequencies recurs (approximately) with period $2k$ in n. In other words, the pattern shows resonance between the time of full radial transit of a ray and the time lapse between arrivals of rays through the separate layers. This is shown schematically in Fig. 8.11 (a).

Again, for a three-layer shell (from (8.4.3))

$$\begin{aligned} S &= (R_1/\pi)\sin(2n\pi T_1/T) - (R_2/\pi)\sin(2n\pi T_3/T) \\ &= (-1)^{n+1}(R_1/\pi)\sin\{(T_2 + T_3 - T_1)n\pi/T\} \\ &\quad - (-1)^{n+1}(R_2/\pi)\sin\{(T_1 + T_2 - T_3)n\pi/T\}. \end{aligned} \tag{8.6.2}$$

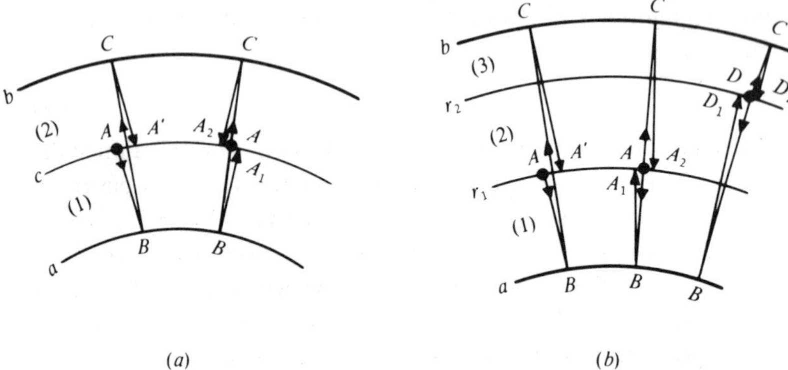

(a) (b)

Fig. 8.11. Resonances which determine the solotone. (a) For two-layer shell. (b) For three-layer shell.

The first term on the right of (8.6.2) arises from resonance between T and the time-lapse between rays through (2) and (3) and through (1), while the second arises from resonance between T and the time-lapse between rays through (1) and (2) and through (3). These paths are shown in Fig. 8.11 (b).

8.6.2 *Constructive interference*

For a more fundamental physical explanation, we turn to equation (8.3.22)

$$\sin(\chi_2 + \chi_1) + R' \sin(\chi_2 - \chi_1) = 0,$$

and re-write it:

$$e^{i(\chi_2+\chi_1)}\{1 - e^{-2i(\chi_2+\chi_1)} + R'e^{-2i\chi_1} - R'e^{-2i\chi_2}\} = 0. \quad (8.6.3)$$

Discarding the factor $\exp i(\chi_2 + \chi_1)$, which cannot vanish, we have

$$1 - e^{-2i\chi_2}\frac{R' + e^{-2i\chi_1}}{1 + R'e^{-2i\chi_1}} = 0. \quad (8.6.4)$$

Using the identity, valid if $|AB| < 1$,

$$\frac{A+B}{1+AB} = A + \frac{B(1-A^2)}{1+AB}$$

$$= A + B(1-A^2)\sum_{j=0}^{\infty}(-AB)^j, \quad (8.6.5)$$

we obtain

$$e^{-2i\chi_2}\{R' + e^{-2i\chi_1}(1 - R'^2)(1 - R'e^{-2i\chi_1} + R'^2 e^{-4i\chi_1} - \ldots)\} = 1. \tag{8.6.6}$$

Now R' is the coefficient of reflection for a wave from region (2) impinging obliquely on the (1, 2) interface, so that the coefficient of transmission is $T' = 1 + R'$. The coefficient of reflection R'' for a wave from region (1) impinging on the interface is $R'' = -R'$, so that

$$T'' = 1 + R'' = 1 - R'. \tag{8.6.7}$$

Hence, for double transmission across the boundary, $T = T'T'' = 1 - R'^2$, and (8.6.6) can be written

$$R'e^{-2i\chi_2} + Te^{-2i(\chi_2+\chi_1)} + TR''e^{-2i(\chi_2+2\chi_1)} + TR''^2 e^{-2i(\chi_2+3\chi_1)} + \ldots = 1. \tag{8.6.8}$$

Here $2\chi_2$ is the phase change in a double transit of the region (2), $2\chi_1$ of the region (1), T is the amplitude factor due to a double crossing of the interface, R'' the amplitude factor at reflection at the interface from region (1). There is no phase or amplitude change on reflection at the surface $r = a$. Equation (8.6.8) states that the vector sum of complex amplitude factors for the infinite series of reflected rays represented there is unity. Thus the rays add up to a wave exactly in phase with whatever original disturbance gave rise to them. The first four of the series of reflected rays are shown in Fig. 8.12. So a value of ω which satisfies the frequency equation (8.6.3) corresponds to a normal mode in which there is constructive interference between all the waves which have undergone multiple reflection at the lower boundary (cf. Officer, 1951).

This is not the only possible pattern of reflections. χ_1 and χ_2 play complementary parts in (8.6.3), so that the roles of regions (1) and (2) in Fig. 8.12 could be reversed.

We note that in this problem the disturbance in one normal mode needs an infinite sum of rays for its representation, just as the disturbance in one pulse needs an infinite sum of normal modes for its representation.

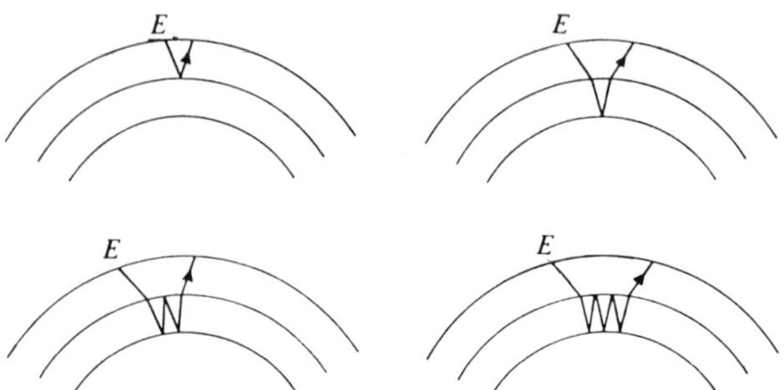

Fig. 8.12. The first four of the sequence of waves multiply reflected – reflected 0, 1, 2, or 3 times at the lower boundary of (1). These are represented by the first four terms on the left-hand side of (8.6.8).

Similar analysis will demonstrate the condition of constructive interference for any number of layers (Lapwood & Sato, 1977).

Nolet & Kennett (1978) have utilised the method of stationary phase (Jeffreys & Jeffreys, 1956, § 17.05) – with regard both to the sum of normal modes over Legendre degree l and to the sum over overtone number n – to derive the asymptotic distribution of eigenfrequencies of toroidal modes of the realistic Earth-model 1066A (Gilbert & Dziewonski, 1975) for large l and n. The model has smoothly varying material constants, and no solotone appears. In a formidable paper, Kennett & Woodhouse (1978) applied asymptotic results for stratified Earth-models (Woodhouse, 1977) to derive changes of phase at discontinuities and in regions of large velocity-gradients. Their results resembled and extended those of § 8.3, and led in the same way to formulae for solotone effects. Their analysis applies to both toroidal and spheroidal oscillations.

The detailed consequences in the frequency-distribution of stratification of the Earth-model are spelled out in a paper by Kennett & Nolet (1979) which deals with toroidal oscillations. In an application to model 1066B (Gilbert & Dziewonski, 1975), which has discontinuities at depths 670 km, 420 km, and sharp changes at the Mohorovičić discontinuity, the frequency distribution for modes up to $l = 700$ and for overtones $n = 1, 2, \ldots, 29$ was found both from asymptotic formulae and by direct computation. There was excellent agreement for $n > 6$. The distribution showed patterns similar to those described in § 8.5.

Few investigations have so far been published on the asymptotic distribution of spheroidal overtones. The reader may consult, in addition to the work of Kennett & Woodhouse mentioned above, papers by Anderssen, Cleary & Dziewonski (1975) and Brodskii & Levshin (1979).

9

EXCITATION OF FREE OSCILLATIONS IN AN ELASTIC SPHERE

We have so far studied the characteristics of free oscillations of a sphere using various models, from a simple uniform elastic sphere to a complicated self-gravitating and rotating realistic Earth, but paying no attention to the initiation of such oscillations. In the actual Earth a large earthquake starts up modes of toroidal and spheroidal oscillation, and these can be identified by Fourier analysis of records of gravimeters and seismometers. In this chapter, we will study the emergence of free oscillations of an elastic sphere following an initial disturbance. However, since the full exploration of the topic is beyond the scope of this book, we will give an outline of the process and place emphasis on matters of physical importance. In this chapter we assume the orthogonality of solutions of the equations of motion for a self-gravitating radially heterogeneous spherical Earth. Woodhouse (1977) obtained beautiful expressions for the orthogonal eigenfunctions. For details, the reader may refer to papers by Backus & Gilbert (1967, 1968, 1970), Saito (1967), Usami, Odaka & Satô (1970), Singh & Ben-Menahem (1969a, b), Alterman & Aboudi (1969), Dziewonski & Gilbert (1972, 1973), Jordan & Anderson (1974), Gilbert & Dziewonski (1975).

9.1 Mathematical formulation of the excitation problem

In this section the excitation of the free oscillations of an elastic sphere is studied in five stages. First we obtain a solution of the equation of motion of a uniform elastic medium which is unbounded and subject to disturbing forces. Secondly, that solution is expressed in terms of spherical polar coordinates with origin at the centre of the sphere. In the third stage we define an *equivalent source function* in terms of discontinuities of stress and displacement. This is used in the fourth stage, at which conditions on the boundary of the sphere are introduced, to obtain a formal solution, in the frequency domain, for the response of the sphere to given applied forces. Finally, that solution in the frequency domain is Fourier-transformed to obtain the solution in the time domain, expressed as a sum of normal modes, each with its appropriate amplitude.

9.1.1 *Displacement in an unbounded medium due to given body forces*

The first stage is to obtain the displacement in an unbounded uniform elastic medium due to a distribution of body forces, following Kelvin (Love, 1927, § 212). By (2.2.6), the vector equation of motion for a uniform medium subject to a body force $\mathbf{F}(\mathbf{r}, t)$ per unit mass is

$$(\lambda + 2\mu)\,\mathrm{grad}\,\mathrm{div}\,\mathbf{u} - \mu\,\mathrm{curl}\,\mathrm{curl}\,\mathbf{u} + \rho\mathbf{F} = \rho\partial_t^2\mathbf{u}. \qquad (9.1.1)$$

By Helmholtz's theorem (see § 2.2), $\mathbf{F}$ and the displacement $\mathbf{u}$ can usually be expressed as

$$\left.\begin{aligned} \mathbf{F} &= \mathrm{grad}\,\Psi + \mathrm{curl}\,\mathbf{A} \qquad (\mathrm{div}\,\mathbf{A} = 0), \\ \mathbf{u} &= \mathrm{grad}\,\Phi + \mathrm{curl}\,\mathbf{B} \qquad (\mathrm{div}\,\mathbf{B} = 0). \end{aligned}\right\} \qquad (9.1.2)$$

Taking divergence and curl of (9.1.1), we find that it is sufficient that

$$\left.\begin{aligned} (\lambda + 2\mu)\nabla^2\Phi + \rho\Psi &= \rho\partial_t^2\Phi, \\ -\mu\,\mathrm{curl}\,\mathrm{curl}\,\mathbf{B} + \rho\mathbf{A} &= \rho\partial_t^2\mathbf{B}. \end{aligned}\right\} \qquad (9.1.3)$$

The solutions of (9.1.3) for Φ and $\mathbf{B}$ at a certain field point $P(\mathbf{r})$, in terms of the known functions Ψ and $\mathbf{A}$, are (Rayleigh, 1896, § 276)

$$\left.\begin{aligned} \Phi &= \frac{1}{4\pi c_{\mathrm{P}}^2}\int_{V''}\frac{1}{R}\Psi(t - R/c_{\mathrm{P}})\,\mathrm{d}V'', \\ \mathbf{B} &= \frac{1}{4\pi c_{\mathrm{S}}^2}\int_{V''}\frac{1}{R}\mathbf{A}(t - R/c_{\mathrm{S}})\,\mathrm{d}V''. \end{aligned}\right\} \qquad (9.1.4)$$

Here the integrals cover the whole of space, including the volume V' in which Ψ and $\mathbf{A}$ are non-zero, and $R = |\mathbf{r}_0 - \mathbf{r}'|$ is the distance to the field-point P from any point Q' in V' (see Fig. 9.1). $\Psi(t - R/c_{\mathrm{P}})$

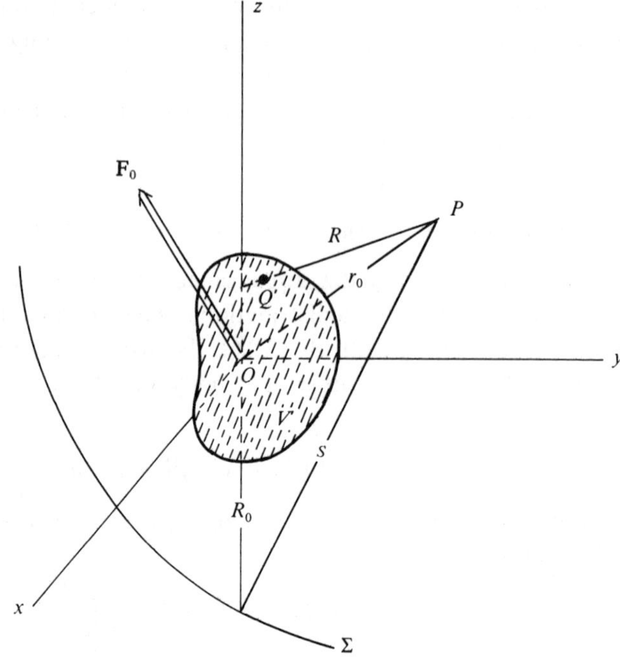

Fig. 9.1. Disturbance at P due to sources in V'.

and $\mathbf{A}(t - R/c_\mathrm{S})$ are the values of Ψ and $\mathbf{A}$ for any point in the whole space V'' at time $t - R/c(c = c_\mathrm{P}$ or $c_\mathrm{S})$. This implies that Ψ and $\mathbf{A}$ are propagated in the medium with velocity c_P and c_S, respectively, to the point P at distance $R = ct'$ for each point Q'' of V''. In other words, the value of Φ at P at time t is the sum of contributions from Ψ at all points of V'' with *retarded time* $t - t'$. Similarly for $\mathbf{B}$.

Now the scalar Ψ and the vector $\mathbf{A}$ from which $\mathbf{F}$ can be derived are given explicitly (Morse & Feshbach, 1953, § 1.5) by

$$\left.\begin{aligned} \Psi &= -\frac{1}{4\pi}\int \frac{1}{R}\,\mathrm{div}'\,\mathbf{F}\,\mathrm{d}V' = \frac{1}{4\pi}\int \mathrm{grad}'\frac{1}{R}\cdot\mathbf{F}\,\mathrm{d}V', \\ \mathbf{A} &= \frac{1}{4\pi}\int \frac{1}{R}\,\mathrm{curl}'\,\mathbf{F}\,\mathrm{d}V' = -\frac{1}{4\pi}\int \mathrm{grad}'\frac{1}{R}\wedge\mathbf{F}\,\mathrm{d}V', \end{aligned}\right\} \tag{9.1.5}$$

on the assumption that $\mathbf{F} = \mathbf{0}$ on the boundary of the volume includ-

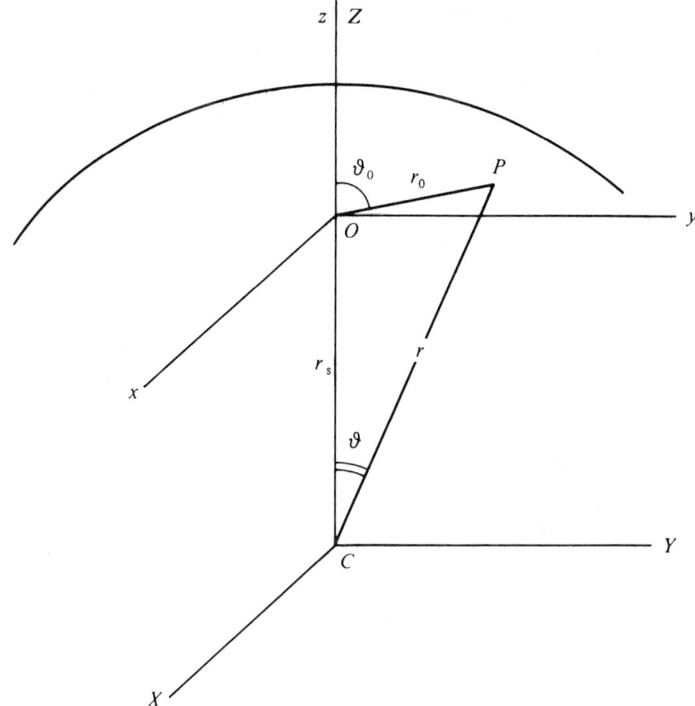

Fig. 9.2. Relation between frames $S_0(Oxyz)$ and $S(CXYZ)$. C is the centre of sphere and O the point of application of concentrated source.

ing V' (for proof, see Appendix § A.6). In (9.1.5) the prime designates an operation with regard to the point $Q'(\mathbf{r}')$ in the volume V'.

We now specialise the disturbing force to a point force concentrated at the origin (Fig. 9.1) with time-variation $f(t)$. So we introduce $\mathbf{F}_0$, defined by

$$\mathbf{F}_0 f(t) = \rho \int \mathbf{F} \mathrm{d}V'. \tag{9.1.6}$$

Then since $\mathrm{grad}'(1/R) = -\,\mathrm{grad}(1/R)$, we have

$$\left.\begin{aligned} \Psi_0 &= -\frac{f(t)}{4\pi\rho}\,\mathrm{grad}\,\frac{1}{R}\cdot\mathbf{F}, \\ \mathbf{A}_0 &= \frac{f(t)}{4\pi\rho}\,\mathrm{grad}\,\frac{1}{R}\wedge\mathbf{F}_0. \end{aligned}\right\} \tag{9.1.7}$$

Now we introduce an appropriate pair of coordinate systems. We use a frame S_0 in which a general point P has coordinates $(r_0, \vartheta_0, \varphi_0)$ with regard to an *origin O at the point of application of the force*, and we reserve (r, ϑ, φ) for coordinates relative to a frame S with origin C at the centre of the sphere (Fig. 9.2). Now in (9.1.7) we may replace R by r_0, since the region V' has been shrunk to the point O. Let us put (9.1.7) with (9.1.4) and employ a formula (due to Gauss) (Love, 1927, § 212) for a surface integral over a sphere Σ centred at P with radius s, viz.

$$\int_\Sigma \operatorname{grad}' \frac{1}{R_0} \mathrm{d}\Sigma = \begin{Bmatrix} 0, & R < s, \\ 4\pi s^2 \operatorname{grad}(1/r_0), & R > s. \end{Bmatrix}. \tag{9.1.8}$$

Here, since we are dealing with the integral over Σ, R of (9.1.7), the distance from the point of application of the force to any point of the whole space, is replaced by R_0 (Fig 9.1), where R_0 and r_0 are the distances from O to any point on Σ and to P, respectively. Then we will have

$$\begin{aligned} \Phi &= -\frac{\mathbf{F}_0}{4\pi c_\mathrm{P}^2} \int_0^\infty \frac{1}{4\pi\rho} f\left(t - \frac{s}{c_\mathrm{P}}\right) \frac{\mathrm{d}s}{s} \int \operatorname{grad} \frac{1}{R_0} \mathrm{d}\Sigma. \\ &= -\frac{1}{4\pi\rho} \mathbf{F}_0 \cdot \operatorname{grad} \frac{1}{r_0} \int_0^{r_0\ c_\mathrm{P}} t' f(t-t') \mathrm{d}t', \\ \mathbf{B} &= \frac{1}{4\pi\rho} \operatorname{grad} \frac{1}{r_0} \wedge \mathbf{F}_0 \int_0^{r_0/c_\mathrm{S}} t' f(t-t') \mathrm{d}t'. \end{aligned} \tag{9.1.9}$$

If

$$f(t) = \exp \mathrm{i}\omega t, \tag{9.1.10}$$

the integral in (9.1.9) becomes

$$\int_0^{r_0/c} t' f(t-t') \mathrm{d}t' = \{-(1/\omega^2) + (1/\omega^2 - r_0/\mathrm{i}c\omega) \exp(-\mathrm{i}\omega r_0/c)\} \exp \mathrm{i}\omega t, \tag{9.1.11}$$

where c takes the value c_P or c_S. The first term of the right-hand side may be neglected since it does not contribute to the wave motion. Thus (9.1.11) may be identified as

$$-\mathrm{i}(r_0^2/c^2) h_1^{(2)}(\omega r_0/c) \exp \mathrm{i}\omega t, \tag{9.1.12}$$

where $h_1^{(2)}(x)$ is the spherical Bessel (Hankel) function of order unity, which may be expressed as

$$h_1^{(2)}(x) = j_1(x) - \mathrm{i}y_1(x) = -x^{-2}(x - \mathrm{i})\exp(-\mathrm{i}x). \quad (9.1.13)$$

Putting (9.1.12) into (9.1.9), we obtain Φ and $\mathbf{B}$, and hence from (9.1.2) we obtain the solution $\mathbf{u}$ of the field equation (9.1.1). The form of the full expression will be determined by that of the applied force $\mathbf{F}$. When $\mathbf{F}$ is a unit force acting in the direction of the z-axis at the origin O (cf. § 7.1), we omit φ_0 because of axial symmetry, and have

$$\left.\begin{aligned} \mathbf{F} &= (\cos\vartheta_0, \ -\sin\vartheta_0, 0)_{S_0}, \\ \operatorname{grad}(1/r_0) &= (-1/r_0^2, \quad 0, 0)_{S_0}, \end{aligned}\right\} \quad (9.1.14)$$

where the suffix to a vector will be used to indicate the reference frame (Fig. 9.2). We now recall that an elementary general time-harmonic solution of the wave equation in polar coordinates has the form $z_l(r)P_l^m(\cos\vartheta)\exp(\mathrm{i}m\varphi + \mathrm{i}\omega t)$, where $z_l(r)$ is a spherical Bessel function and $P_l^m(\cos\vartheta)$ the Associated Legendre function, l, m being integers with $m \leq l$. So we cast our formula for the field due to a point source into this form. Remembering that $\cos\vartheta_0 = P_1^0(\cos\vartheta_0)$, we obtain

$$\left.\begin{aligned} \Phi &= -\frac{\mathrm{i}}{4\pi\rho c_P^2} P_1^0(\cos\vartheta_0) h_1^{(2)}\left(\frac{\omega r_0}{c_P}\right) \exp \mathrm{i}\omega t, \\ \mathbf{B} &= -\frac{\mathrm{i}}{4\pi\rho c_S^2} \operatorname{curl}\left\{ r_0 P_1^0(\cos\vartheta_0) h_1^{(2)}\left(\frac{\omega r_0}{c_S}\right), 0, 0 \right\}_{S_0} \exp \mathrm{i}\omega t. \end{aligned}\right\} \quad (9.1.15)$$

Force systems capable of simulating actual earthquakes are usually assumed to be double couples. The field of a double couple can be obtained from the field due to a point source $\mathbf{F}$ by applying a suitable differential operator (Love, 1927, § 132). Since derivatives of Associated Legendre functions are also Associated Legendre functions, the scalar potential for a double couple can be expressed by a sum of terms containing $P_l^m(\cos\vartheta_0)h_m^{(2)}(\omega r_0/c)$, where l, m take values $(0, 0)$, $(1, 0)$, $(2, 0)$, $(1, 1)$, $(2, 1)$, $(2, 2)$, and the coefficients depend on the orientation of the double couple. The vector potential can be similarly expressed by the curl of a vector whose transverse components are zero and whose radial component is a sum of terms containing

$$P_l^m(\cos \vartheta_0) h_m^{(2)}(\omega r_0 / c_S).$$

9.1.2 *Field due to a point force expressed in spheroidal polar coordinates*

In the first stage, the origin of the coordinate system $(r_0, \vartheta_0, \varphi_0)$ in frame S_0 was taken at the point O where the concentrated force was applied. In the second stage, we will use spherical polar coordinates (r, ϑ, φ) whose origin C is at the centre of the elastic sphere, and examine the field due to a point force at $(r_s, 0, 0)$. The relation between these two coordinate systems is shown in Fig. 9.2. There is symmetry about the axis CZ, so that φ_0, φ do not appear as variables. Onda & Sato (1969) have given the formula which relates our functions of (r_0, ϑ_0) to eigenfunctions expressed in terms of (r, ϑ) for the sphere, viz.

$$h_l^{(2)}(kr_0) P_l^m(\cos \vartheta_0) = \begin{cases} \sum_{i=0}^{\infty} \varepsilon_i^{l,m} j_{i+m}(kr) P_{i+m}^m(\cos \vartheta), & r_s > r, \\ \sum_{i=0}^{\infty} \zeta_i^{l,m} h_{i+m}^{(2)}(kr) P_{i+m}^m(\cos \vartheta), & r_s < r, \end{cases} \tag{9.1.16}$$

where the coefficients $\varepsilon_i^{l,m}$ are

$$\left. \begin{aligned} \varepsilon_i^{0,0} &= (2i+1) h_i^{(2)}(kr_s), \\ \varepsilon_i^{1,0} &= (2i+1)[\{i h_i^{(2)}(kr_s)/(kr_s)\} - h_{i+1}^{(2)}(kr_s)], \\ \varepsilon_i^{2,0} &= \tfrac{3}{2}(2i+1)[-\tfrac{2}{3} h_i^{(2)}(kr_s) + \{2 h_{i+1}^{(2)}(kr_s)/(kr_s)\} \\ &\quad + \{i(i-1) h_i^{(2)}(kr_s)/(kr_s)^2\}], \\ \varepsilon_i^{1,1} &= (2i+3) h_{i+1}^{(2)}(kr_s)/(kr_s), \\ \varepsilon_i^{2,1} &= 3(2i+3)[\{i h_{i+1}^{(2)}(kr_s)/(kr_s)^2\} \\ &\quad - \{h_{i+2}^{(2)}(kr_s)/(kr_s)\}], \\ \varepsilon_i^{2,2} &= 3(2i+5) h_{i+2}^{(2)}(kr_s)/(kr_s)^2, \end{aligned} \right\} \tag{9.1.17}$$

and $\zeta_i^{l,m}$ can be obtained from $\varepsilon_i^{l,m}$ by replacing $h_{i+m}^{(2)}$ by j_{i+m}.

We now recall that an elementary general solution of the wave equation in polar coordinates with the origin at the centre of the sphere has the form $z_l(hr)P_l^m(\cos\vartheta)\,{}^{\cos}_{\sin}\,m\varphi\exp\mathrm{i}\omega t$ (2.2.23). In order to be able to deal with boundary conditions on the surface of the sphere, we transform our expression for the field due to a point source into this form, using formula (9.1.16). $\varepsilon_i^{l,m}$ and $\zeta_i^{l,m}$ are constant coefficients.

For a unit force in the z-direction, Φ of (9.1.15) reduces to

$$\Phi = \{-\mathrm{i}/(4\pi\rho c_{\mathrm{P}}^2)\}\exp(\mathrm{i}\omega t)\Lambda(c_{\mathrm{P}}), \tag{9.1.18}$$

where, for $r > r_{\mathrm{s}}$,

$$\Lambda(c) = \sum_{i=0}^{\infty} \zeta_i^{1,0}\, h_i^{(2)}(\omega r/c) P_l(\cos\vartheta). \tag{9.1.19}$$

To obtain $\mathbf{B}$, we write

$$\begin{aligned}
\operatorname{curl}\{F,0,0\}_{S_0} &= \operatorname{curl}\{F\cos(\vartheta_0-\vartheta), F\sin(\vartheta_0-\vartheta), 0\}_{\Gamma_{\mathrm{p}}},\\
&= \operatorname{curl}\left\{\frac{F}{r_0}(r - r_{\mathrm{s}}\cos\vartheta), \frac{F}{r_0}r_{\mathrm{s}}\sin\vartheta, 0\right\}_{\Gamma_{\mathrm{p}}},\\
&= \operatorname{curl}\left\{\frac{F}{r_0}r, 0, 0\right\}_{\Gamma_{\mathrm{p}}}\\
&\quad + \operatorname{curl}\left\{-\frac{F}{r_0}r_{\mathrm{s}}\cos\vartheta, \frac{F}{r_0}r_{\mathrm{s}}\sin\vartheta, 0\right\}_{\Gamma_{\mathrm{p}}},\\
&= \operatorname{curl}\left\{\frac{F}{r_0}r, 0, 0\right\}_{\Gamma_{\mathrm{p}}} + \operatorname{curl}\left\{0, 0, -\frac{F}{r_0}r_{\mathrm{s}}\right\}_{\Gamma_{\mathrm{C}}}
\end{aligned} \tag{9.1.20}$$

where the suffixes Γ_{p} and Γ_{C} indicate respectively the polar coordinate frame S and the rectangular Cartesian frame at the centre C of the sphere, as shown in Fig. 9.2. Then

$$\begin{aligned}
\mathbf{B} = \{-\mathrm{i}/(4\pi\rho c_{\mathrm{S}}^2)\}\exp\mathrm{i}\omega t\,[&\operatorname{curl}\{r\Lambda(c_{\mathrm{S}}), 0, 0\}_{\Gamma_{\mathrm{p}}}\\
&+ \operatorname{curl}\{0, 0, -r_{\mathrm{s}}\Lambda(c_{\mathrm{S}})\}_{\Gamma_{\mathrm{C}}}].
\end{aligned}$$

For the region $r < r_{\mathrm{s}}$, the appropriate $\Lambda(c)$ will be obtained by changing $\zeta_i^{l,m}$ to $\varepsilon_i^{l,m}$ and $h_i^{(2)}$ to j_i.

9.1.3 *Conditions on source surface equivalent to applied force*

It proves to be difficult to solve directly the *inhomogeneous* wave equation (containing a term representing the applied force) for a heterogeneous elastic sphere. We therefore introduce the concept of an 'equivalent source function' as follows. A point source on the spherical surface $r = r_s$ is usually replaceable by discontinuities of the displacements and stress components on that surface. Denoting discontinuities at the 'source surface' $r = r_s$ by Δ, we define Δ by

$$\Delta D = \lim_{\varepsilon \to 0} (D_{r=r_s+\varepsilon} - D_{r=r_s-\varepsilon}), \tag{9.1.21}$$

where D stands for any displacement or surface stress component. Taking the particular source which is a time-harmonic unit point force at O in the z-direction and given by Φ only, the discontinuities on the source surface $r = r_s$, derived from (9.1.18) and (9.1.19), are

$$\left.\begin{array}{l} \Delta u = 0, \qquad \Delta v = 0, \qquad \Delta w = 0, \\ \Delta\widehat{rr} = -(1/4\pi r_s^2)\exp(i\omega t)\sum\limits_{l=0}^{\infty}(2l+1)P_l(\cos\vartheta), \quad \Delta\widehat{r\vartheta} = 0, \quad \Delta\widehat{r\varphi} = 0. \end{array}\right\} \tag{9.1.22}$$

The expression for $\Delta\widehat{rr}$ in (9.1.22) diverges for $\vartheta = 0$, and is to be understood according to the conventions of the theory of distributions. Discontinuities which accompany a point force exist only at the point of application of that force, displacements and surface stress being continuous at all other points.

For each pair of values of l and m, however, displacement and stress components on the source surface show discontinuities. Let us call the combination of all these discontinuities on the source surface the *equivalent source function*, defined so that the solutions of the following two problems are the same:

(i) Express the disturbing force in polar coordinates (r, ϑ, φ) referred to the centre of the sphere, and solve the inhomogeneous wave equations for the sphere with the given disturbing force and the appropriate surface and centre conditions.

(ii) Divide the sphere into two parts by the source surface. Seek suitable fields satisfying the homogeneous wave equations in the two parts and match these fields so as (a) to satisfy the surface and centre conditions and (b) to give exactly the discontinuities of displacement and stress which form the equivalent source function at the surface.

In this chapter we use the second method to obtain the solution to an excitation problem due to a body force, adopting an equivalent source function. It is to be expected that a body force will usually excite both toroidal and spheroidal oscillations.

9.1.4 *Solutions in the frequency domain*

In the fourth stage we proceed to find analytic solutions of the problem. It was shown in Chapters 5 and 6 that solutions of the homogeneous wave equations in a radially heterogeneous and self-gravitating elastic sphere are obtainable by integrating the differential equations (5.2.4) for spheroidal oscillations and (6.1.19) for toroidal oscillations. When we introduce the equivalent source function we can use the same homogeneous differential equations for the problem of excitation.

First consider a source with time variation $\exp i\omega t$. Let $\delta_1^{lm}, \delta_2^{lm}, \ldots, \delta_6^{lm}$ be discontinuities at the source surface (factors containing t, ϑ and φ being omitted) corresponding to terms of parameters (l, m) in

$$\Delta\widehat{r\varphi},\quad \Delta W;\quad \Delta\widehat{rr},\quad \Delta\widehat{r\vartheta},\quad \Delta U,\quad \Delta V, \tag{9.1.23}$$

respectively. Thus δ_1 and δ_2 refer to toroidal motion, $\delta_3, \ldots, \delta_6$ to spheroidal. We introduce the notation

$$\left.\begin{aligned} E_l^{\mathrm{t}} &= \widehat{r\varphi} = \mu(\mathrm{d}W_l/\mathrm{d}r - W_l/r), \\ E_l^{\mathrm{S}} &= \widehat{rr} = (\lambda + 2\mu)\mathrm{d}U_l/\mathrm{d}r + 2\lambda U_l/r - l(l+1)\lambda V_l/r, \\ E_l^{\mathrm{T}} &= \widehat{r\vartheta} = \mu(\mathrm{d}V_l/\mathrm{d}r - V_l/r + U_l/r). \end{aligned}\right\} \tag{9.1.24}$$

Then for toroidal motion of the type $\{\bar{u}(\omega), \bar{v}(\omega), \bar{w}(\omega)\} \exp i\omega t$, the solution providing the given discontinuities on the source surface has

$$\begin{Bmatrix} \bar{u}(\omega) \\ \\ \bar{v}(\omega) \\ \\ \bar{w}(\omega) \end{Bmatrix} = -\left(\frac{r_{\mathrm{s}}}{a}\right)^2 \sum_{l,m} W_l(r) \frac{\delta_1^{lm} W_l(r_{\mathrm{s}}) - \delta_2^{lm} E_l^{\mathrm{t}}(r_{\mathrm{s}})}{E_l^t(a)} \begin{Bmatrix} 0 \\ \\ \dfrac{m}{\sin\vartheta} P_l^m(\cos\vartheta)\,{}^{(-\sin)}_{\ \ \cos}\, m\varphi \\ \\ -\partial_\vartheta P_l^m(\cos\vartheta)\,{}^{\cos}_{\sin}\, m\varphi \end{Bmatrix}, \tag{9.1.25}$$

where $W_l(r)$ is a solution of the ordinary differential equation for W_l (6.1.19) which arises from separation of variables; in a uniform sphere it would be a spherical Bessel function. On the right-hand side of (9.1.25), W_l and E_l^{t} are functions of the angular frequency ω.

Similarly, for spheroidal motion a time-harmonic solution of the type $\{\bar{u}(\omega), \bar{v}(\omega), \bar{w}(\omega)\} \exp \mathrm{i}\omega t$ has

$$\begin{Bmatrix} \bar{u}(\omega) \\ \bar{v}(\omega) \\ \bar{w}(\omega) \end{Bmatrix} = -\left(\frac{r_{\mathrm{s}}}{a}\right)^2 \sum_{l,m} \frac{\delta_3^{lm} U_l(r_{\mathrm{s}}) + l(l+1)\delta_4^{lm} V_l(r_{\mathrm{s}}) - \delta_5^{lm} E_l^{\mathrm{s}}(r_{\mathrm{s}}) - l(l+1)\delta_6^{lm} E_l^{\mathrm{T}}(r^{\mathrm{s}})}{E_l^{\mathrm{S}}(a)}$$
$$\times \begin{Bmatrix} U_l(r) P_l^m(\cos\vartheta) {}^{\cos}_{\sin} m\varphi \\ V_l(r)\partial_\vartheta P_l^m(\cos\vartheta) {}^{\cos}_{\sin} m\varphi \\ V_l(r) \dfrac{m}{\sin\vartheta} P_l^m(\cos\vartheta) {}^{(-\sin)}_{\cos} m\varphi \end{Bmatrix}, \tag{9.1.26}$$

where $U_l(r)$, $V_l(r)$ are solutions of ordinary differential equations arising after separation of variables. On the right-hand side of (9.1.26), U_l, V_l, E_l^{S} and E_l^{T}, include ω as parameter. In (9.1.25) and (9.1.26), W_l, U_l and V_l satisfy the differential equation and the condition at the centre, but do not satisfy the conditions at the free boundary. Solutions (9.1.25) and (9.1.26) lie in the frequency domain, and are to be multiplied by a factor $\exp \mathrm{i}\omega t$.

9.1.5 *Solutions in the time domain*

In the final stage we proceed to obtain solutions in the time domain. To obtain the response to a disturbing force of time-variation $f(t)$, we must write

$$f(t) = \frac{1}{2\pi} \int_{-\infty}^{\infty} \bar{f}(\omega) \mathrm{e}^{\mathrm{i}\omega t} \, \mathrm{d}\omega, \tag{9.1.27}$$

where $\bar{f}(\omega)$ is the Fourier transform of $f(t)$, and convolve the solution (9.1.25) or (9.1.26) with $\bar{f}(\omega)$ to obtain the response to $f(t)$, namely

$$\begin{Bmatrix} u(t) \\ v(t) \\ w(t) \end{Bmatrix} = \frac{1}{2\pi} \int_{-\infty}^{\infty} \begin{Bmatrix} \bar{u}(\omega) \\ \bar{v}(\omega) \\ \bar{w}(\omega) \end{Bmatrix} \bar{f}(\omega) \mathrm{e}^{\mathrm{i}\omega t} \, \mathrm{d}\omega. \tag{9.1.28}$$

If we evaluate (9.1.28) by contour integration, we can show that the integral reduces to the sum of all contributions from residues at poles (that is, at the zeroes of the denominator) on the positive real axis of ω. The denominators in (9.1.25) and (9.1.26) are in fact stress components on the surface, and their zeroes correspond to the eigenfrequencies ${}_n\omega_l$ of toroidal and spheroidal oscillations. (Contributions from poles of $\bar{f}(\omega)$ give small terms independent of time; they do not contribute to oscillatory motion. We omit them here.) Thus we obtain finally from (9.1.25) (toroidal motion)

$$\begin{Bmatrix} u(t) \\ v(t) \\ w(t) \end{Bmatrix} = -\mathrm{i}\left(\frac{r_s}{a}\right)^2 \sum_{l,m,n} \left[\left\{ \frac{\bar{f}(\omega)\mathrm{e}^{\mathrm{i}\omega t}}{\partial_\omega E_l^{\mathrm{t}}(a)} (\delta_1^{lm} W_l(r_s) - \delta_2^{lm} E_l^{\mathrm{t}}(r_s)) W_l(r) \right\}_{\omega = {}_n\omega_l} \times \begin{Bmatrix} 0 \\ \dfrac{m}{\sin\vartheta} P_l^m(\cos\vartheta) \begin{pmatrix} -\sin \\ \cos \end{pmatrix} m\varphi \\ -\partial_\vartheta P_l^m(\cos\vartheta) \begin{matrix} \cos \\ \sin \end{matrix} m\varphi \end{Bmatrix} \right], \tag{9.1.29}$$

and from (9.1.26) (spheroidal motion)

$$\begin{Bmatrix} u(t) \\ v(t) \\ w(t) \end{Bmatrix} = -\mathrm{i}\left(\frac{r_s}{a}\right)^2 \sum_{l,m,n} \left[\left[\left\{ \frac{\bar{f}(\omega)\mathrm{e}^{\mathrm{i}\omega t}}{\partial_\omega E_l^{\mathrm{S}}(a)} (\delta_3^{lm} U_l(r_s) + l(l+1)\delta_4^{lm} V_l(r_s) - \delta_5^{lm} E_l^{\mathrm{S}}(r_s) - l(l+1)\delta_6^{lm} E_l^{\mathrm{T}}(r_s)) \times \begin{Bmatrix} U_l(r) \\ V_l(r) \\ V_l(r) \end{Bmatrix} \right\}_{\omega = {}_n\omega_l} \times \begin{Bmatrix} P_l^m(\cos\vartheta) \begin{matrix} \cos \\ \sin \end{matrix} m\varphi \\ \partial_\vartheta P_l^m(\cos\vartheta) \begin{matrix} \cos \\ \sin \end{matrix} m\varphi \\ \dfrac{m}{\sin\vartheta} P_l^m(\cos\vartheta) \begin{pmatrix} -\sin \\ \cos \end{pmatrix} m\varphi \end{Bmatrix} \right] \right]. \tag{9.1.30}$$

In (9.1.29) and (9.1.30) the functions U_l, V_l, W_l are eigenfunctions and E_l^{S}, E_l^{T} and E_l^{t} are corresponding stress components. $\bar{f}(\omega)$ has been defined as the Fourier transform of the time function of the disturbing force, and the notation $\sum_{l,m,n}$ indicates summation for all possible combinations of l, m and n. Thus the solution of the equations of motion of an elastic sphere when a body force is applied is shown to be expressed by a weighted sum of the free toroidal and spheroidal oscillations. In our example (9.1.22), when the applied force is a unit force pointing in the z-direction, all δ_i^{lm} vanish except δ_3^{lm} (cf. (9.1.23)). Hence no toroidal oscillation is excited and the solution is given by (9.1.30) with δ_4^{lm}, δ_5^{lm} and δ_6^{lm} all zero.

The quantity $\{\partial_\omega E_l^{\mathrm{t}}(a)\}_{\omega={}_n\omega_l}$, which is introduced when we compute the residue at the pole ${}_n\omega_l$, is obtained approximately as

$$\frac{\{E_l^{\mathrm{t}}(a)\}_{\omega=\omega_+} - \{E_l^{\mathrm{t}}(a)\}_{\omega=\omega_-}}{\omega_+ - \omega_-}, \tag{9.1.31}$$

where ω_+ and ω_- are close to the corresponding eigenfrequency ${}_n\omega_l$ and on opposite sides of ${}_n\omega_l$. $\{\partial_\omega E_l^{\mathrm{S}}(a)\}_{\omega={}_n\omega_l}$ is treated in the same way. It is more convenient in numerical work to attribute small changes $\pm\varepsilon_E$ to $E_l^{\mathrm{t}}(a)$ and then to calculate the values ω_+ and ω_- which satisfy the regularity condition at the centre of the sphere. The quantities $\{\partial_\omega E_l^{\mathrm{t}}(a)\}_{\omega={}_n\omega_l}$ and $\{\partial_\omega E_l^{\mathrm{S}}(a)\}_{\omega={}_n\omega_l}$ may also be found from the kinetic energy of toroidal and spheroidal modes integrated over the whole sphere (Saito, 1967, Singh & Ben-Menahem, 1969a, b). Such methods of evaluation are better because they do not depend on numerical differentiation, which generally has lower accuracy than numerical integration. Employing these numerical values of $\{\partial_\omega E_l^{\mathrm{t}}(a)\}_{\omega={}_n\omega_l}$ and $\{\partial_\omega E_l^{\mathrm{S}}(a)\}_{\omega={}_n\omega_l}$ with the eigenfunctions corresponding to the eigenfrequency ${}_n\omega_l$ and the Fourier transform $\bar{f}(\omega)$ of the time function of the disturbing force, we can calculate the displacements at points of the elastic sphere due to the point force at radius r_s by (9.1.29) and (9.1.30) as functions of time. Such calculated displacements are *theoretical seismograms*.

In (9.1.29) and (9.1.30) the factor $(r_s/a)^2$ is a constant depending on the depth of the source. The first factor in the bracket, $\bar{f}(\omega) \exp i\omega t$ depends on the time function of the applied force. U_l, V_l, W_l, E_l^{t}, E_l^{S}, E_l^{T}, $\partial_\omega E_l^{\mathrm{t}}$ and $\partial_\omega E_l^{\mathrm{S}}$ are determined when the structure of the elastic sphere is given. $\delta_1^{lm}, \ldots, \delta_6^{lm}$ are discontinuities depending on the type of applied force. The last group of factors, containing r, ϑ and φ, are functions of the position of the point of observation. Thus in (9.1.29) and (9.1.30) the factors separate into groups dependent on (a) the characteristics of the applied force, (b) the whole structure of the sphere and (c) the position of the observation point.

9.2 Examples of theoretical seismograms

In this section we show examples of theoretical seismograms of disturbances for which there is axial symmetry and $m=0$. In all of them we graph displacement components on the surface as functions of time, the unit of time (abscissa) being the time for an S wave to encircle the sphere. The unit of displacement (ordinate) is arbitrary. We use the brackets ( ) and [ ] with the following meanings:

$$({}_n u_l) = \sum_{j=0}^{n} \sum_{i=0}^{l} {}_j u_i ; \qquad [{}_j u_l] = \sum_{i=0}^{l} {}_j u_i , \tag{9.2.1}$$

where ${}_j u_i$ is the term in (9.1.30) which pertains to $l=i, n=j, m=0$. Similarly for transverse displacements v and w. We assume that readers already possess a basic knowledge of seismological concepts – such as phase-names and travel-time curves (Bullen, 1959). In the following figures we use the usual notation for reflected phases. In addition we use: + and − after the phase-name to indicate the pulses which have travelled along the major and minor arcs respectively (e.g. SS +, SS−); numerals to indicate repetitions (e.g. 4S for SSSS and 2ScS for ScSScS); G_{2n+1} to indicate a shear wave which has circled the Earth n times and also travelled the shorter arc between epicentre and station once; G_{2n+2} similarly has travelled n circuits and the longer arc once.

9.2.1 *Toroidal oscillations set up in a uniform shell by a transient torque*

Fig. 9.3 (Usami & Satô, 1972) shows the theoretical seismogram of the azimuthal displacement w for toroidal oscillations of a uniform mantle enclosing a liquid core. Axial symmetry is assumed, so that $m=0$. We take $b/a = 0.54474$ (cf. § 3.1). The external force consists of the stress component $\widehat{r\varphi}$ applied at the surface on a small annulus $\vartheta_1 \le \vartheta \le \vartheta_2$. It is a 'boxcar-shaped' function of time. Thus

$$\widehat{r\varphi} = \Phi(\vartheta, \varphi) f(t), \tag{9.2.2}$$

where

$$\Phi(\vartheta, \varphi) = \left\{ \begin{array}{ll} 1 & \vartheta_1 \le \vartheta \le \vartheta_2, \\ 0 & \vartheta < \vartheta_1 \text{ and } \vartheta > \vartheta_2, \end{array} \right\} \tag{9.2.3}$$

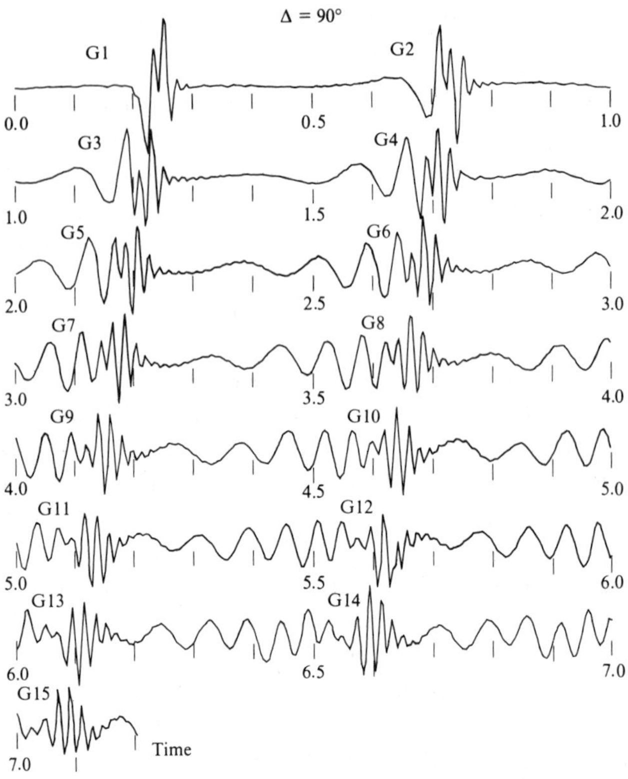

Fig. 9.3. Theoretical seismogram of w at $\Delta = 90^\circ$ for toroidal oscillations following a boxcar-shaped torsional impulse on an annulus. 60 fundamental modes ($n = 0, l = 1, 2, \ldots, 60$) are summed. Unit of time: $2\pi a/c_S$. Surface waves G are easily identified. (From Usami & Satô, 1972.)

and

$$f(t) = \begin{cases} 1/2t_1 & -t_1 \le t \le t_1, \\ 0 & t < -t, \text{ and } t > t_1. \end{cases} \tag{9.2.4}$$

Since this applies a pure twist around the z-axis, only toroidal oscillations are excited. For numerical computations, we take

$$\vartheta_1 = 0.02 \text{ rad}, \qquad \vartheta_2 = 0.04 \text{ rad}, \qquad t_1 = 0.015 \text{ units}, \tag{9.2.5}$$

the unit of time being $2\pi a/c_S$. Fig. 9.3 represents the surface value at $\Delta = 90^\circ$ of $({}_0w_{60})$, that is, the sum of contributions from fundamental modes ($n = 0$) with Legendre degree $l = 1$ to 60 (modes with $l = 0$ are

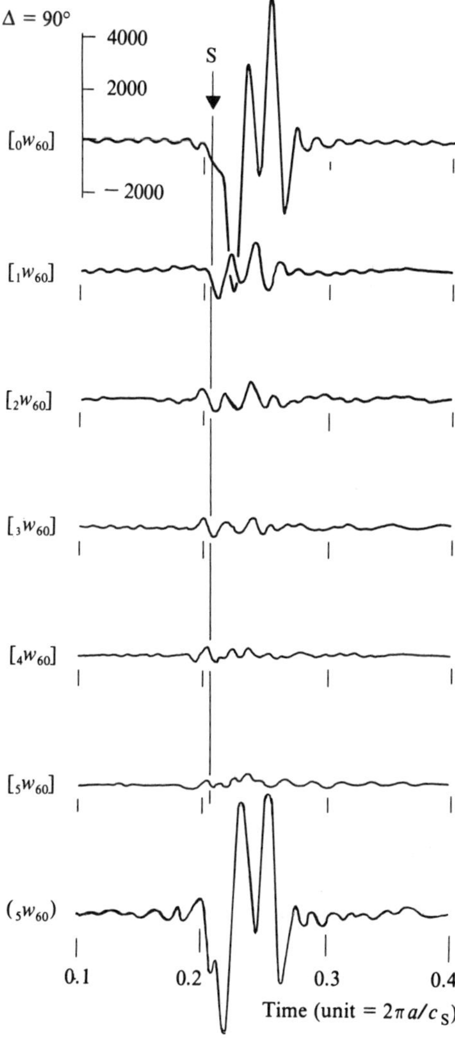

Fig. 9.4. Theoretical seismograms of toroidal oscillations for a uniform sphere with perfect-liquid core. The contributions of radial overtones summed over degree $l'([_nw_{60}] = \sum_{l'=1}^{60} {_nw_{l'}})$ are plotted as functions of time at $\Delta = 90°$ for $n = 0, 1, 2, 3, 4, 5$. The lowest graph shows the sum of all of these. The arrow indicates the theoretical arrival time of S; some of the radial overtones seem to start before S, but their sum does not. (From Satô *et al.*, 1963.)

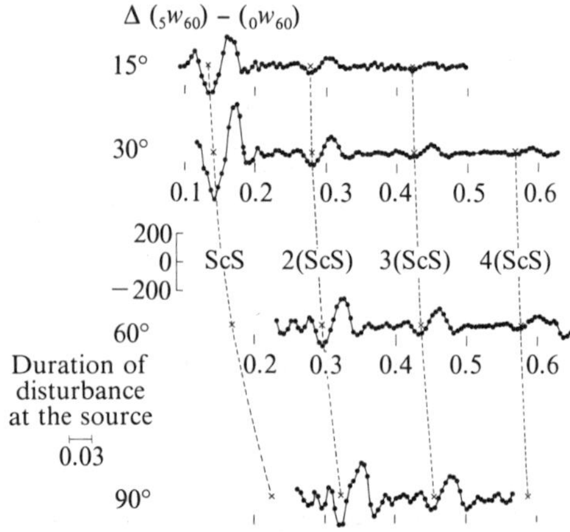

Fig. 9.5. Displacement w due to ScS. The sum of the first five overtones $((_5w_{60})-(_0w_{60}))$ is plotted as a function of time for the same model as in Fig. 9.3. (×) indicates the theoretical arrival time of nScS. (From Satô *et al.*, 1963.)

omitted). Since the source has axial symmetry, the theoretical seismogram is independent of φ. In the figure we see surface waves propagated along the minor arc (G1) and major arc (G2), and those which have circled the sphere repeatedly (G3, G5, ... , G4, G6, ...). The dispersion of the waves is remarkable: the minimum group-velocity of dispersive waves is seen to be slightly greater than c_S.

Fig. 9.4 (Satô *et al.*, 1963) shows other theoretical seismograms for the same spherical model as in Fig. 9.3. Fig. 9.4 shows for the first half-unit of time the summed contributions for $l = 1$ to 60 ($l = 0$ is omitted) at $\Delta = 90°$ for the fundamental ($n = 0$) and for the first five overtones. S marks the theoretical arrival time of the direct S wave. Contributions from some of the overtones seem to begin before the arrival of the direct S wave, but it appears from the lowest graph (in Fig. 9.4) that these apparently earlier arrivals cancel each other when superposed. In this example the contributions from fundamental modes have greater amplitude than those from overtones, so that in the superposed seismograms the overtones are almost concealed in the fundamental modes.

Fig. 9.5 shows the role of the overtones in producing reflected

S pulses. The structure of the elastic sphere, and the applied external force, are the same as in Fig. 9.3. Contributions of five overtones only ($n = 1, 2, \ldots, 5$) are summed; the resultant sum shows 2ScS and 3ScS clearly. Since the ordinate scales in Fig. 9.4 and Fig. 9.5 are consistent, we realise that in this example core-reflected S waves would be concealed in a theoretical seismogram in which the contributions from the fundamental modes were included. These figures strongly suggest that the overtones make the main contribution to the body wave (see also Fig. 9.7).

Fig. 9.6 shows the first arrival of $({}_5w_{60})(l = 0$ omitted) for a uniform shell with a perfect-liquid core. The external force system and structure of the model are the same as for Fig. 9.3. The core creates a shadow zone for the direct S wave at $\Delta > \Delta_c = 113.987°$, and the S wave is diffracted into this shadow zone. In each part of Fig. 9.6, the onset of the disturbance, determined by eye as for observed seismograms, is shown by an arrow. The time indicated by the arrow fits well the travel-time calculated simply by geometrical optical theory. It can be seen that in the region beyond $\Delta = 112.5°$, diffracted waves precede the arrival of the ray reflected from the surface. Thus it becomes clear that the difference between non-dimensional frequencies for a uniform sphere and a uniform shell, which is small for large values of l and n, results in a remarkable difference in theoretical seismograms, namely in the appearance for the uniform shell of the ScS wave and the wave diffracted by the core. From the standpoint of geometrical ray theory and Huygens' principle, these features are expected. From the standpoint of eigenfrequency analysis, diffracted waves correspond to a maximum in the group-velocity curve, the value of which is close to the group-velocity of a wave grazing the core (Fig. 3.2.).

9.2.2 *Spheroidal oscillations set up in a gravitating Earth-model by transient normal stresses on a cap*

We now give a second example (Usami & Satô, 1966) in which the overtones make the main contribution to the body wave, and the fundamentals to the surface wave. Here we assume a homogeneous sphere, but we allow for self-gravity in spheroidal oscillations. Fig. 9.7 is the theoretical seismogram for spheroidal motion of

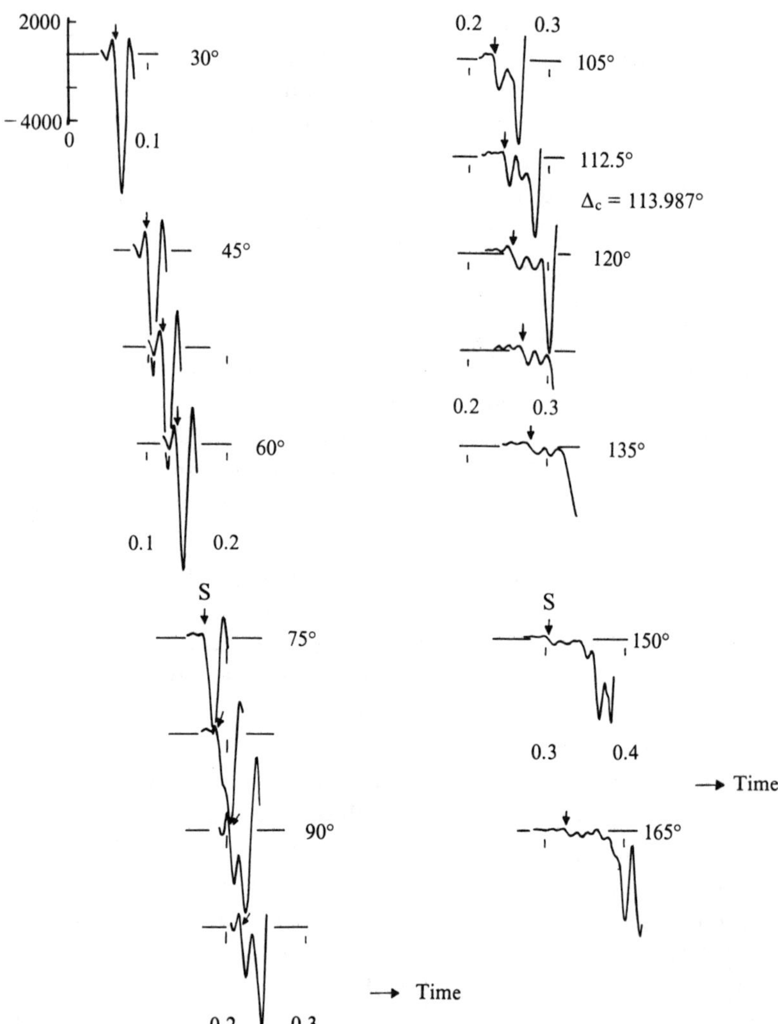

Fig. 9.6. Displacement w near the time of first motion (indicated by arrow) for the same model as in Fig. 9.3. This shows diffracted waves in the shadow zone $\Delta > \Delta_c$, where $\Delta_c \approx 114°$. (From Satô *et al.*, 1963.)

this uniform self-gravitating sphere near the antipole. The disturbing force, due to normal stress on a small disc on the surface, is in the form of a single square wave:

$$\widehat{rr} = \Phi(\vartheta, \varphi) f(t), \tag{9.2.6}$$

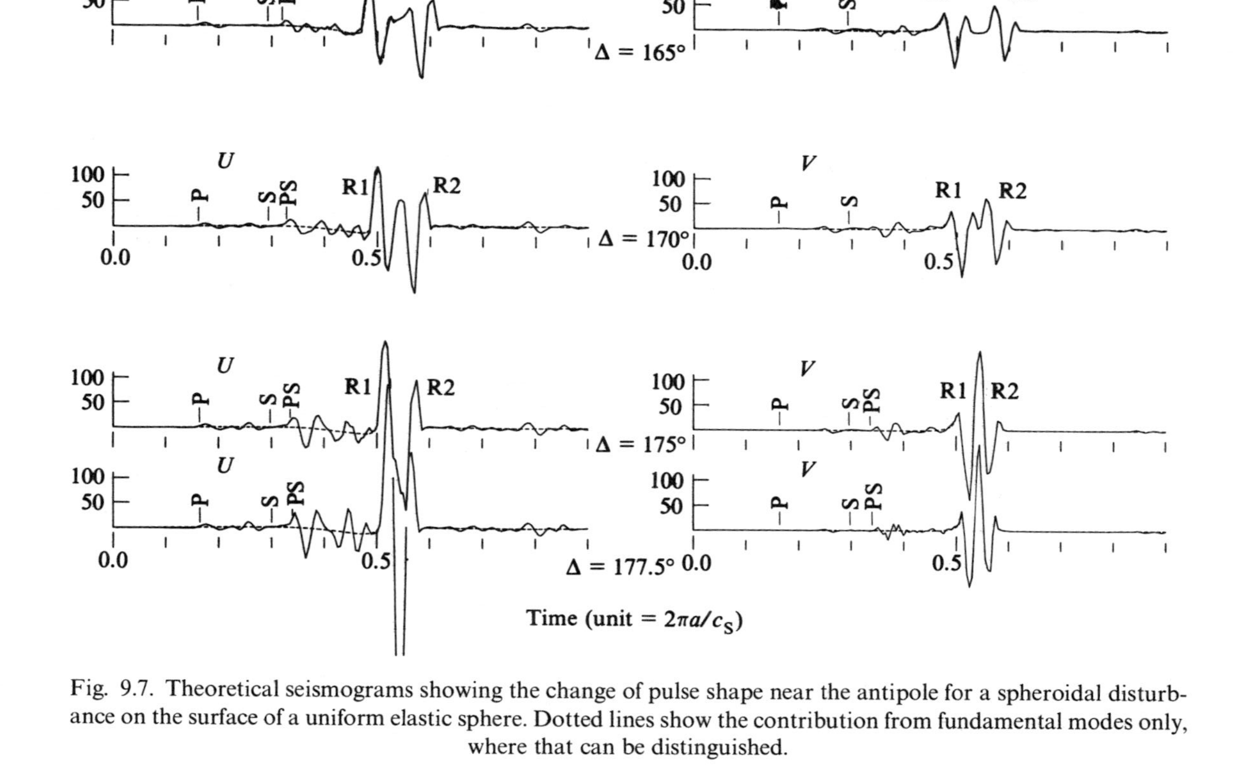

Fig. 9.7. Theoretical seismograms showing the change of pulse shape near the antipole for a spheroidal disturbance on the surface of a uniform elastic sphere. Dotted lines show the contribution from fundamental modes only, where that can be distinguished.

where

$$\Phi(\vartheta, \varphi) = \begin{cases} 1 & \vartheta \le \vartheta_0, \\ 0 & \vartheta > \vartheta_0, \end{cases} \tag{9.2.8}$$

$$f(t) = \begin{cases} -1 & -t_1 \le t < 0, \\ 1 & 0 < t < t_1, \\ 0 & |t| > t_1. \end{cases} \tag{9.2.8}$$

We take

$$\vartheta_0 = 0.04 \text{ rad}, \qquad t_1 = 0.02 \text{ units}. \tag{9.2.9}$$

The unit of time is $2\pi a / c_S$. Contributions are summed over $n = 0, 1, \ldots, 9$ and $l = 1, 2, \ldots, l_{max}$, l_{max} being taken as follows:

$$\left.\begin{array}{lcccccccccc} n = & 0 & 1 & 2 & 3 & 4 & 5 & 6 & 7 & 8 & 9 \\ l_{max} = & 160 & 79 & 73 & 69 & 64 & 60 & 60 & 39 & 44 & 44 \end{array}\right\} \tag{9.2.10}$$

(In order to get good representation of surface waves we take many modes for $n = 0$.) In Fig. 9.7, solid lines represent the sum of contributions from all the modes listed in (9.2.10), while dotted lines show the contribution from the fundamental modes only. In the passage of surface waves (indicated by R1 and R2), contributions from overtones are negligibly small compared with those from fundamental modes. Contrariwise overtones make the main contribution to the body waves. Since the surface waves (R1 and R2) and the body wave PS converge (geometrically) towards the antipole, their amplitude increases rapidly as they approach the antipole. By contrast, P waves which arrive directly from the origin through the core do not show any increase in amplitude near the antipole.

Fig. 9.8 (Usami, Satô, Landisman & Odaka, 1968) sets out theoretical seismograms for spheroidal motion of a self-gravitating sphere, again consisting of a uniform mantle with a perfect-liquid core. The applied external force is that given by (9.2.6) to (9.2.8), except that we adopt values

$$\vartheta_0 = 0.012 \text{ rad}, \qquad t_1 = 0.004 \text{ units}, \tag{9.2.11}$$

corresponding to a sharper and more concentrated pulse.

The figure exhibits the sum of contributions from overtones only. The maximum Legendre degree of those modes employed in

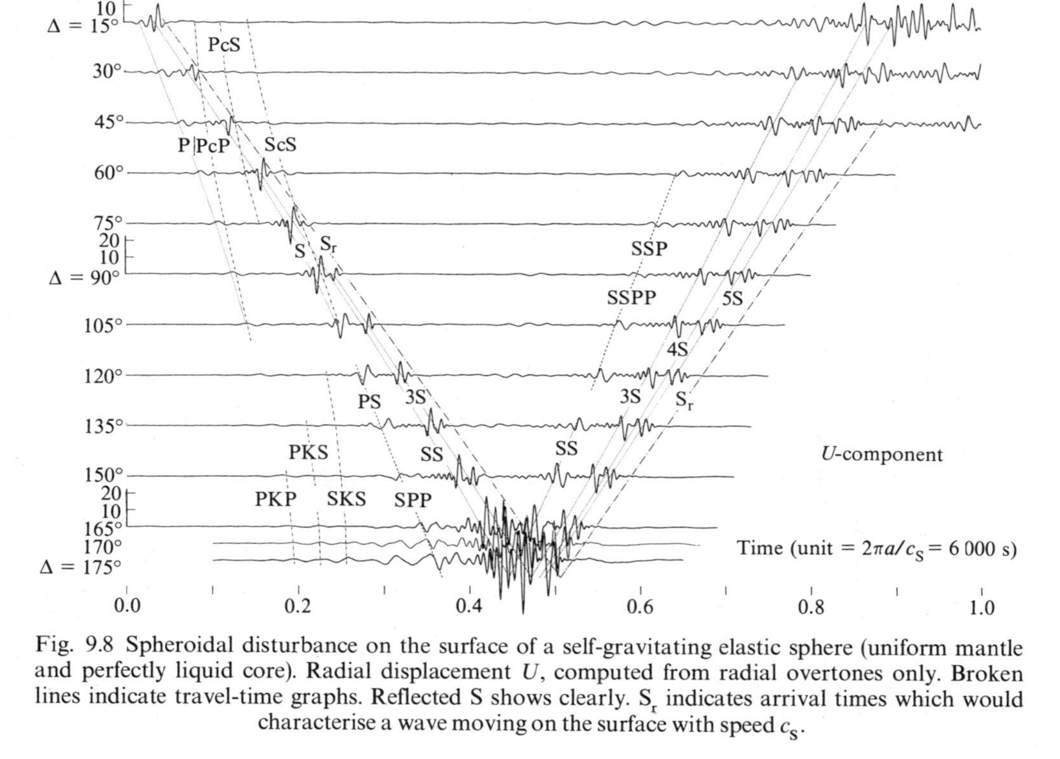

Fig. 9.8 Spheroidal disturbance on the surface of a self-gravitating elastic sphere (uniform mantle and perfectly liquid core). Radial displacement U, computed from radial overtones only. Broken lines indicate travel-time graphs. Reflected S shows clearly. S_r indicates arrival times which would characterise a wave moving on the surface with speed c_S.

the computation for each n (except n_s) is l_{max},† given by:

$$\left.\begin{array}{llllllllll} n = n_s & 1 & 2 & 3 & 4 & 5 & 6 & 7 & 8 \\ l_{max} = 37 & 481 & 470 & 460 & 452 & 445 & 439 & 432 & 429 \end{array}\right\} \quad (9.2.12)$$

Travel-time curves calculated by geometrical ray theory are shown in the figure by various broken lines carrying the phase-names. S_r indicates the arrival time of a pulse moving along the surface with S-wave velocity, if such a pulse existed. The direct P and S waves, waves diffracted at the surface, waves reflected at and refracted by the core can be identified in these theoretical seismograms.

Theoretical seismograms such as those shown above lead us to expect that:

(i) The whole disturbance in an elastic sphere may be expressed as the sum of contributions from all modes of free toroidal and spheroidal oscillations.
(ii) The wave patterns in the theoretical seismogram may show more and more resemblance to those seen in records of natural earthquakes, as the number of modes summed increases and as the Earth-model approaches the structure of the real Earth.
(iii) A general correspondence between fundamental modes and surface waves, and between overtone modes and body waves may be valid for any elastic sphere.

9.3 The stress glut and source moment tensor

In earlier work on excitation of an elastic medium by a local disturbance, the simplest source models were employed. For example, a point source of dilatational waves was used to represent an explosion, and a point source of rotational waves (the limit of a twist applied on the surface of an embedded sphere) to generate shear waves. The disturbance due to a doublet was then derived by differentiation of that due to a point source (Love, 1927, § 213), and that due to a multipole by further differentiation and superposition.

†For each n (except for $n = n_s$) we sum over frequencies up to a given bound (dimensionless frequency $\eta = 520$), that is, over modes with period more than 11.5 s. l_{max} corresponds to that bound for each n.

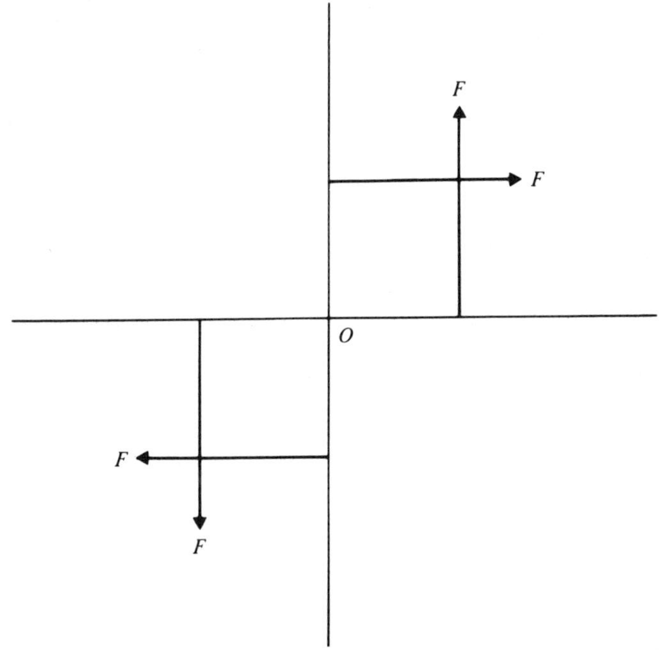

Fig. 9.9. Double couple without moment.

The point source described as 'double couple without moment', which is the limit of the pattern of equal forces shown in Fig 9.9, proved to have a radiation pattern similar to that of many earthquakes, and has been extensively employed as a seismic source model.

Burridge & Knopoff showed in a classic paper (1964) that the radiation pattern from a double couple without moment is identical with that produced by a point tangential *dislocation* appropriately orientated in an infinite elastic space. This result was obtained as a particular case of the general relations which they proved to hold between the radiation fields of dislocations and equivalent force systems. Dahlen (1972) extended the results of Burridge & Knopoff to apply to the fields of sources in a stratified sphere with given prestress. The response of a layered elastic sphere to a harmonic point source had been derived by Ben-Menahem (1964) and Saito (1967), and the response to a finite dislocation by Singh & Ben-Menahem (1969a, b).

In § 9.1 we showed how to deal with a source which is confined to a spherical surface of given radius concentric with a radially heterogeneous sphere. We now proceed to consider a more general source, following the recent paper of Backus & Mulcahy (1976) and using the modern and more general representation of a source by a moment tensor. We have made use of the elegant unpublished treatment of Woodhouse (1977).

9.3.1 *The stress glut tensor and equivalent body force density*

We consider a medium which is perfectly elastic but anisotropic (obeying a generalised Hooke's Law) except for the *source region*, V_s, in which the physical stress tensor σ_{ij} (actual stress) differs from the value τ_{ij} given by Hooke's Law from the actual strain. Inside V_s, let

$$\tau_{ij} - \sigma_{ij} = \Gamma_{ij}. \tag{9.3.1}$$

By the usual examination of an elementary tetrahedron we find that τ_{ij} is symmetric, while σ_{ij} is symmetric from Hooke's Law and the symmetry of the strain tensor. Γ_{ij} is therefore a symmetric tensor. It was called by Backus & Mulcahy the *stress glut*. We may define the source region as that volume in which the stress glut is non-zero.

We apply this concept in a non-rotating self-gravitating body of volume V, which before the application of the source had local density ρ_0, gravitational potential ψ_0 and hydrostatic stress $-p_0\delta_{ij}$ in equilibrium. Now suppose that after the source begins to disturb the body the local displacement is u_i, while the density and potential become $\rho_0 + \rho, \psi_0 + \psi$. We assume that there is no body force except gravity. Then the linearised equations of momentum and gravitation are:

$$\left.\begin{aligned} \rho_0 \partial_t^2 u_i &= \partial_j \sigma_{ij} - (\rho_0 + \rho)\partial_i(\psi_0 + \psi), \\ \nabla^2(\psi_0 + \psi) &= -4\pi G(\rho_0 + \rho), \\ \rho &= -\partial_k(\rho_0 u_k). \end{aligned}\right\} \tag{9.3.2}$$

Outside V_s

$$\sigma_{ij} = \tau_{ij} = -(p_0 - u_k \partial_k p_0)\delta_{ij} + c_{ijkl}\partial_l u_k, \tag{9.3.3}$$

where c_{ijkl} are the usual elastic constants; but inside V_s

$$\sigma_{ij} = \tau_{ij} - \Gamma_{ij},$$

so that the first equation of (9.3.2) is

$$\rho_0 \partial_t^2 u_i = \partial_j \tau_{ij} - (\rho_0 + \rho)\partial_i(\psi_0 + \psi) - \partial_j \Gamma_{ij}. \tag{9.3.4}$$

The extra term $-\partial_j \Gamma_{ij}$ shows that the existence of the stress glut tensor has the same effect as an extra body force density $f_i = -\partial_j \Gamma_{ij}$. The vector $\mathbf{f}$ is called the *equivalent body force density*.

If S is a closed surface just outside V_s,

$$\begin{aligned}\int_{V_s} f_i \mathrm{d}V &= -\int_{V_s} \partial_j \Gamma_{ij} \mathrm{d}V \\ &= -\int_S \Gamma_{ij} n_j \mathrm{d}S \\ &= 0,\end{aligned} \tag{9.3.5}$$

and

$$\begin{aligned}\int_{V_s} (\mathbf{x} \wedge \mathbf{f})_i \mathrm{d}V &= -\varepsilon_{ijk} \int_{V_s} x_j \partial_l \Gamma_{kl} \mathrm{d}V \\ &= \varepsilon_{ijk} \left(\int_{V_s} \delta_{jl} \Gamma_{kl} \mathrm{d}V - \int_S x_j \Gamma_{kl} n_l \mathrm{d}S \right) \\ &= 0,\end{aligned} \tag{9.3.6}$$

since Γ_{kj} is symmetric and $\Gamma_{kl} = 0$ on S. Thus the net force and net torque on the body, due to the stress glut or equivalent body force density, are zero.

We define also a *moment tensor* $\mathbf{M}$ by

$$\begin{aligned}M_{ij}(t) &= \int_{V_s} x_i f_j(\mathbf{x}, t) \mathrm{d}V \\ &= -\int_{V_s} x_i \partial_k \Gamma_{jk} \mathrm{d}V \\ &= -\int_{V_s} \partial_k (x_i \Gamma_{jk}) \mathrm{d}V + \int_{V_s} \Gamma_{ji} \mathrm{d}V \\ &= \int_{V_s} \Gamma_{ji} \mathrm{d}V,\end{aligned} \tag{9.3.7}$$

by Gauss's theorem. Thus $M_{ij}(t)$ is symmetric.

9.3.2 *Disturbance due to source of equivalent body force density* $\mathbf{f}(\mathbf{x}, t)$

In a non-rotating self-gravitating layered sphere of volume V, the displacement $\mathbf{u}(\mathbf{x}, t)$ due to body force $\mathbf{f}(\mathbf{x}, t)$ is found by solving

$$\mathbf{L}(\mathbf{u}) + \mathbf{f} = \rho_0 \partial_t^2 \mathbf{u} \qquad \text{in } V, \tag{9.3.8}$$

where

$$L_i(\mathbf{u}) = \rho_0 \{u_{k,k}\psi_{0,i} - \psi_{,i} - (u_k \psi_{0,k})_{,i}\} + (c_{ijkl} u_{l,k})_{,j}, \tag{9.3.9}$$
$$(\psi_{,i} \equiv \partial_i \psi),$$

with the conditions $[u_i]_-^+ = 0$,

$$[c_{ijkl} u_{l,k}]_-^+ = 0 \text{ at each layer boundary,} \tag{9.3.10}$$

ensuring continuous displacements and balanced stresses.

We solve in terms of eigenfunctions. Let the suffix k denote the mode (q, l, m, n), where $q = 1$ means 'toroidal', $q = 2$ 'spheroidal', l is Legendre degree, m azimuthal order and n overtone number. Let $\mathbf{u}_k(\mathbf{x}, \omega_k)$ be the eigenfunction corresponding to the eigenvalue (angular frequency) ω_k, so that

$$\mathbf{L}(\mathbf{u}_k) + \rho_0 \omega_k^2 \mathbf{u}_k = 0. \tag{9.3.11}$$

It can be shown that the operator $\mathbf{L}$ is self-adjoint, so that the set of eigenfunctions is complete, and we can postulate a solution of (9.3.8)

$$\mathbf{u}(\mathbf{x}, t) = \sum_k a_k(t) \mathbf{u}_k(\mathbf{x}), \tag{9.3.12}$$

the sum being taken over all normal modes. When the eigenfunctions are orthonormal

$$\int_V \rho_0 \mathbf{u}_j^*(\mathbf{x}) \cdot \mathbf{u}_k(\mathbf{x}) \mathrm{d}V = \delta_{jk}, \tag{9.3.13}$$

(*) denoting complex conjugate; for δ_{jk} to be non-zero, the set (q, l, m, n) for j must be identical with that for k.

Substituting $\mathbf{u}$ of (9.3.12) into (9.3.8) and using (9.3.11) we have

$$-\sum_k \rho_0 \omega_k^2 a_k \mathbf{u}_k + \mathbf{f} = \sum_k \rho_0 (\mathrm{d}^2 a_k / \mathrm{d}t^2) \mathbf{u}_k. \tag{9.3.14}$$

We take the scalar product of each side of (9.3.14) with $\mathbf{u}_j^*$ and integrate throughout V to obtain

$$\begin{aligned} \mathrm{d}^2 a_j/\mathrm{d}t^2 + \omega_j^2 a_j &= \int_{V_s} \mathbf{u}_j^* \cdot \mathbf{f} \mathrm{d}V \\ &= F_j(t), \end{aligned} \tag{9.3.15}$$

defining $F_j(t)$. The solution of this ordinary differential equation (for instance by variation of parameters) depends on $F_j(t)$ through

$$a_j(t) = \frac{1}{\omega_j} \int_{-\infty}^{t} \sin \omega_j(t-\tau) F_j(\tau) \mathrm{d}\tau. \tag{9.3.16}$$

If the source region is small compared with the wave-length corresponding to ω_j, we may introduce the first-order expansion

$$\mathbf{u}_j(\mathbf{x}) = \mathbf{u}_j(\mathbf{x}_0) + (x_i - x_{0i})\{\partial_i \mathbf{u}_j(\mathbf{x})\}_{\mathbf{x}=\mathbf{x}_0} + O(2) \tag{9.3.17}$$

into the definition of F_j, and obtain, using (9.3.5), the first-order approximation

$$F_j(t) = (\partial_i u_{jk}^*)_{\mathbf{x}=\mathbf{x}_0} M_{ik}(t), \tag{9.3.18}$$

where

$$M_{ik}(t) = \int_{V_s} x_i f_k(\mathbf{x}, t) \mathrm{d}V, \tag{9.3.19}$$

as in (9.3.7). Thus we observe that the source moment tensor enters naturally when we retain only first-order terms in the expansion of $\mathbf{u}_j(\mathbf{x})$ about $\mathbf{x}_0$.

Since the strain tensor $\mathbf{E}_j$ in the jth mode is

$$\begin{aligned} e_{ik}^j &= \tfrac{1}{2}(u_{jk,i} + u_{ji,k}), \\ (\partial_i u_{jk}^*)_{\mathbf{x}=\mathbf{x}_0} M_{ik} &= e_{ik}^{j*}(\mathbf{x}_0) M_{ik}, \end{aligned}$$

so that

$$F_j(t) = \mathbf{E}_j^*(\mathbf{x}_0) : \mathbf{M}(t), \tag{9.3.20}$$

where (:) indicates double contraction. Since the main characteristic of an earthquake is the sudden change in stress at the source, it is more appropriate, unless we are concerned with residual stresses, to use the *moment rate tensor* $\mathscr{M}(t) = \mathrm{d}\mathbf{M}/\mathrm{d}t$ rather than $\mathbf{M}$, and to replace (9.3.20) by

$$\dot{F}_j(t) = \mathbf{E}_j^*(\mathbf{x}_0) : \mathscr{M}(t).$$

9.3.3 *Spectral stacking and stripping*

When we are faced with observations of an earthquake, such as observed values of displacement $\mathbf{u}(\mathbf{x}_i, t)$ at a set of N stations $\mathbf{x}_i$ $(i = 1, 2, \ldots, N)$ scattered over the surface of the Earth, two related problems present themselves. We want to discover from the observations two sets of quantities, (a) parameters describing the inner structure of the Earth: $\rho(r)$, $\lambda(r)$, $\mu(r)$, and possibly measures of anelasticity, as functions of depth, and (b) parameters describing the character of the earthquake source: location, time of occurrence and source moment tensor, dislocation pattern or equivalent force density distribution. Gilbert & Dziewonski (1975) have set out a clear description of a powerful iterative procedure by which, if we start from a model based on plausible assumptions as to Earth structure and source mechanism, the data can be handled so as to improve the model step by step. As will emerge, the process is best carried out in the frequency domain rather than in the time domain.

From (9.3.12) and (9.3.16), the theoretical response to source distribution $\mathbf{f}(\mathbf{x}, t)$ is

$$\mathbf{u}(\mathbf{x}, t) = \sum_k \mathbf{u}_k(\mathbf{x}) \int_{-\infty}^{t} \frac{1}{\omega_k} \sin \omega_k (t - \tau) F_k(\tau) \mathrm{d}\tau, \qquad (9.3.21)$$

where $F_j(t)$ is defined in (9.3.15). Integrating by parts so as to introduce the moment rate tensor, we get

$$\begin{aligned} \mathbf{u}(\mathbf{x}, t) &= \sum_k \mathbf{u}_k(\mathbf{x}) \int_{-\infty}^{t} \frac{1}{\omega_k^2} \{1 - \cos \omega_k (t - \tau)\} \frac{\mathrm{d}F_k}{\mathrm{d}\tau} \mathrm{d}\tau \\ &= \sum_k \mathbf{u}_k(\mathbf{x}) \mathbf{E}_k^* : \mathscr{M}(t) * C_k(t), \end{aligned} \qquad (9.3.22)$$

where

$$C_k(t) = (1 - \cos \omega_k t) H(t), \qquad (9.3.23)$$

and $(*)$ indicates convolution. (Gilbert & Dziewonski, in order to allow for attenuation, introduced the parameter $\alpha_k = \frac{1}{2}\omega_k / Q_k$, Q_k being the quality factor for vibrations of angular frequency ω_k (see § 10.3). $C_k(t)$ then took the form

$$\{1 - \exp(-\alpha_k t) \cos \omega_k t\} H(t).)$$

Transferring the problem into the frequency domain, we denote the Fourier transform of $\phi(t)$ by

$$\bar{\phi}(\omega) = \int_{-\infty}^{\infty} \phi(t)\exp(-\mathrm{i}\omega t)\mathrm{d}t.$$

The transform of (9.3.22) is

$$\begin{aligned} \bar{\mathbf{u}}(\mathbf{x}, \omega) &= \sum_k \omega_k^{-2} \bar{\mathbf{u}}_k(\mathbf{x}) \mathbf{E}_k^*(\mathbf{x}_0) : \bar{\mathscr{M}}(\omega) \bar{C}_k(\omega) \\ &= \sum_k \mathbf{b}_k(\mathbf{x}, \omega) \bar{C}_k(\omega), \end{aligned} \tag{9.3.24}$$

where

$$\mathbf{b}_k = \omega_k^{-2} \bar{\mathbf{u}}_k(\mathbf{x}) \mathbf{E}_k^*(\mathbf{x}_0) : \bar{\mathscr{M}}(\omega).$$

This has an obvious advantage over the expression in (9.3.22) in that the convolution has been replaced by a simple product. $\mathbf{u}(\mathbf{x}, t)$ is the theoretical displacement, computed from the assumed *starting model*. Our observations give, instead of $\mathbf{u}, \mathbf{u}'(\mathbf{x}_i, t)$ measured at seismographic stations.

From the starting model we can compute provisional eigenfrequencies ω_k and the corresponding displacements $\mathbf{u}_k(\mathbf{x})$ in normal modes. From the assumed source mechanism we can compute $\mathbf{E}_k^*(\mathbf{x}_0)$ and $\bar{\mathscr{M}}(\omega)$. We could then replace the theoretical quantities $\bar{\mathbf{u}}(\mathbf{x}, \omega)$ by the observed spectra $\bar{\mathbf{u}}'(\mathbf{x}_i, \omega)$, and, seeking now an improved set of eigenfrequencies, regard the equations

$$\bar{\mathbf{u}}'(\mathbf{x}_i, \omega) = \sum_k \mathbf{b}_k(\mathbf{x}_i, \omega) \bar{C}_k'(\omega) \qquad (i = 1, 2, \ldots, N)$$

as a set of N equations for $\bar{C}_k'(\omega)$, which has replaced $\bar{C}_k(\omega)$. We prefer to adopt, however, the elegant method of Gilbert & Dziewonski, which picks out each mode in turn by an appropriate weighting of observations. It is based on earlier work of Gilbert & Backus (1965).

We return to (9.3.24) and recall that k denotes the set (q, l, m, n). We will disregard the splitting effect of rotation or departure from axial symmetry, and deal with the degenerate system in which ω_k is the same for all m when (q, l, n) are given. Then in (9.3.24) we can perform the summation over m, and, indicating the set (q, l, n) by p, obtain

$$\bar{\mathbf{u}}(\mathbf{x}, \omega) = \sum_p \mathbf{a}_p(\mathbf{x}, \omega) \bar{C}_p(\omega), \tag{9.3.25}$$

where

$$\mathbf{a}_p = \omega_p^{-2} \sum_m \bar{\mathbf{u}}_{p,m} \mathbf{E}^*_{p,m}(\mathbf{x}_0) : \bar{\mathscr{M}}(\omega). \tag{9.3.26}$$

We now call on the orthogonal properties of normal modes. We form the scalar product of each side of (9.3.25) with $\mathbf{a}^*_p$ and integrate over the unit sphere Ω.

It can be shown that, if p' is (q', l', n'),

$$\int \mathbf{a}^*_p \cdot \mathbf{a}_{p'} \, d\Omega = A^{ql}_{nn'} \delta_{qq'} \delta_{ll'}, \tag{9.3.27}$$

where

$$A^{ql}_{nn'} = \int \mathbf{a}^*_{qln} \cdot \mathbf{a}_{qln'} \, d\Omega, \tag{9.3.28}$$

that is, the only terms which survive in (9.3.27) are those for which $q' = q, l' = l$. So from (9.3.25) we get

$$\begin{aligned} v_{qln}(\omega) &= \int \bar{\mathbf{u}}(\mathbf{x}, \omega) \cdot \mathbf{a}^*_{qln} \, d\Omega \\ &= \sum_{n'} A^{ql}_{nn'} \bar{C}_{qln'}(\omega), \end{aligned} \tag{9.3.29}$$

where $C_{qln'}$ represents $C_{p'}$ of (9.3.25). $v_{qln}(\omega)$ is the spectrum for the multiplet (q, l, n) obtained by scalar-multiplying $\mathbf{u}(\mathbf{x})$ at each point of the unit sphere by the complex conjugate of the vector $\mathbf{a}_{qln'}(\mathbf{x}, \omega)$ in (9.3.26), which is derived from the eigenfunctions of the multiplet (q, l, n) and the parameters of the source.

Corresponding to $v_{qln}(\omega)$, we can derive from the observations

$$v'_{qln}(\omega) = \sum_i \bar{\mathbf{u}}'(\mathbf{x}_i, \omega) \cdot \mathbf{a}^*_{qln}. \tag{9.3.30}$$

Here $\bar{\mathbf{u}}'(\mathbf{x}_i, \omega)$ is the Fourier transform, that is the spectrum, derived from observations $\mathbf{u}'(\mathbf{x}_i, t)$; (9.3.30) thus represents the weighted sum of derived spectra, the weighting being determined by mode and source. v'_{qln} is called the *spectral stack*. The method of construction means that in the stack ω_{qln} will be the dominant frequency, others being suppressed by destructive interference in the *stacking* process.

We can go a step further in extracting ω_{qln}. Corresponding to (9.3.29), v'_{qln} is given by

$$v'_{qln} = \sum_{n'} A^{ql}_{nn'} \bar{C}'_{qln'},$$

or

$$\mathbf{v}'_{ql} = \mathbf{A}^{ql} \bar{\mathbf{C}}'_{ql},$$

where $\mathbf{v}$ and $\bar{\mathbf{C}}$ are column vectors and $\mathbf{A}$ is a matrix. Inverting the matrix we get

$$\bar{\mathbf{C}}'_{ql}(\omega) = (\mathbf{A}^{ql})^{-1} \mathbf{v}_{ql}.$$

The element $\bar{C}'_{qln}(\omega)$ is the spectrum derived from the observations (and the model) for the frequency of the (q, l, n) modes. By the inversion of the matrix we have been able to pick out an estimate of a single frequency. The process is called *stripping* by Gilbert & Dziewonski (1975) who have demonstrated its effectiveness in identifying eigenfrequencies, especially those of overtones.

These improved estimates of eigenfrequencies can be used to construct a refined Earth model. Gilbert & Dziewonski go on to show how to improve the model of source mechanism. In the paper quoted above they use their refined models to show that for two deep earthquakes the source moment tensor (a) had a compressive isotropic part, and (b) had principal axes which rotated during the earthquake. Sailor & Dziewonski (1978) have applied the techniques of stacking and stripping in their discussion of normal mode attenuation.

10

RECENT AND CURRENT RESEARCH

In this chapter we give a brief account of recent work on free oscillations of the Earth. The Earth-models described in previous chapters were simple ones, but were nevertheless sufficient to enable us to portray and explain the main characteristics of the Earth's toroidal and spheroidal oscillations. The present task of theoretical seismologists is to refine those models.

When we enquire which features of the Earth have been excluded from our models, three present themselves immediately as candidates for investigation – lateral heterogeneity, pre-stress together with anisotropy, and anelasticity. These form the topics of later sections of this chapter; but before starting to study them we will attack a more general subject of very great importance. This is the analysis of *inverse problems*.

10.1 Inverse problems

So far we have posed *direct* problems. Assuming that we know the functions $\lambda(r)$, $\mu(r)$ and $\rho(r)$, that is, the distribution with depth of the relevant physical constants of the material of the Earth, we have been able to set up field equations and boundary conditions; solving these we have obtained the eigenfrequencies uniquely. But our real problems are not of this type. The eigenfrequencies are in fact observed, and $\lambda(r)$, $\mu(r)$ and $\rho(r)$ are not. Thus the true problem of seismology is not the direct problem but the *inverse* problem – given the observations, to derive the material parameters as functions of radius.

To pose the problem is to discover immediately that there is no unique solution. For however good our observations, they produce only a finite number of units of information. But through a finite number of points can be passed an infinite number of continuous curves.

Analogously, $\lambda(r)$, $\mu(r)$ and $\rho(r)$ cannot be uniquely determined. We naturally enquire whether it is possible to state criteria which discriminate among the infinite family of possible solutions. So we embark on the study of the methods of inversion.

Inversion to obtain an Earth-model may take, as given material, any set of geophysical data, such as travel-times, dispersion curves, measurements of attenuation; but we will work in terms of eigenfrequencies of free oscillations as our observed quantities. We wish to find an Earth-model, or rather a set of Earth-models, which fit the data well. It will be necessary to design a method for constructing the set, and criteria for testing the fit and discriminating between models.

Each method of construction sets out from an assumed model – one which experience has shown to be a good first approximation, and which may have been chosen with the help of evidence outside the data-set which we aim to fit. For instance, aiming to fit eigenfrequencies, we may start from a Jeffreys–Bullen model obtained by studies of travel-times, and some physical hypothesis as to density distribution. We then need a process – probably iterative – of refinement of the model by adjusting parameters. Such processes are described in an excellent review article by Parker (1977). See also Sabatier (1977).

We must distinguish between linear inversion, where the observed quantities are linear functionals of unknown functions such as $\lambda(r)$, $\mu(r)$, $\rho(r)$, and non-linear inversion. The former is much more accessible to analysis; and even non-linear problems are often best attacked by local linearisation.

When a suitable model has been found we will need ways of describing its quality, namely the precision and resolution of the estimates which it provides of the parameters we have sought. Our next section describes an attack on this problem which has already become classical.

10.1.1 *Gross Earth functionals*

In the years following 1967, Backus & Gilbert (1967, 1968, 1970), working together at the University of California, San Diego, produced an elegant and powerful general technique for treating the linear inverse problem. Dahlen (1974) developed a beautiful method of constructing linear inversion formulae for the fine structures of eigenfrequencies. In this section we follow the ideas of Backus & Gilbert as expounded by Burridge (1976). In order to simplify the discussion we deal with an artificial one-dimensional example given by Gilbert (1971). We consider a perfect sphere of radius unity, with physical constants independent of ϑ and φ, and for which all physical parameters are known as functions of radius, except the density $\rho(r)$. We denote by $\mathscr{M}$ the space of all possible Earth-models m_i, each of which corresponds to a particular density distribution $\rho(r, m_i)$, $0 \leq r \leq 1$. We assume that one of these models, m_{E}, corresponds to the actual Earth, with density $\rho(r, m_{\mathrm{E}})$.

Suppose further that we have information about P *gross Earth functionals*

$$\gamma_p, \quad p = 1, 2, \ldots, P$$

(such as mass, moment of inertia, eigenfrequency of a particular mode, of the whole Earth), and that the theoretical formula by which γ_p is calculated from $\rho(r)$ is

$$\gamma_p = \int_0^1 G_p(r)\rho(r)\mathrm{d}r. \tag{10.1.1}$$

$G_p(r)$ is called the *data kernel* corresponding to γ_p; it is assumed to be known from direct theory.

10.1.2 *Error-free data*

First we assume that the *observed* value of γ_p is $\bar{\gamma}_p$, precisely known and free from observational error. $\bar{\gamma}_p$ we call the *gross Earth datum.* Clearly the P quantities $\bar{\gamma}_p$ are insufficient to determine $\rho(r)$ in $0 \leq r \leq 1$. But let us consider how to make best use of the observations.

We specify a point $r = r_0$, and seek a linear combination of data-kernels

$$A(r, r_0) = \sum_1^P a_p G_p, \tag{10.1.2}$$

which will make

$$\rho_{\mathrm{a}}(r_0, m) \equiv \int_0^1 A(r, r_0)\rho(r, m)\mathrm{d}r \tag{10.1.3}$$

a good estimate of $\rho(r_0, m_E)$. Since $\bar{\gamma}_p$ are assumed free from error, we have

$$\begin{aligned}\rho_a(r_0, m_E) &= \sum_1^P \int_0^1 a_p G_p(r)\rho(r, m_E)\,dr \\ &= \sum_1^P a_p \gamma_p(m_E) \\ &= \sum_1^P a_p \bar{\gamma}_p, \end{aligned} \tag{10.1.4}$$

being functions of r_0.

How shall a_p be chosen? If $A(r, r_0)$ were actually $\delta(r - r_0)$, $\rho_a(r_0)$ would be exactly $\rho(r_0, m_E)$, but A must be a linear combination of data-kernels, so that all we can do is make it as delta-like as possible, given G_p. The criterion for this property of A can be chosen as we wish; Backus & Gilbert (1968) opted for the minimisation of the *spread*

$$S(r_0) = 12 \int_0^1 A^2(r - r_0)^2\,dr. \tag{10.1.5}$$

It can be proved that, when a_p are so chosen, $A(r, r_0)$ resembles $\delta(r - r_0)$ in concentrating non-negligible values of the integrand in $\int_0^1 A(r, r_0)\rho(r)\,dr$ into the neighbourhood of r_0. But we must quickly remark that S as defined is positive definite and takes a minimum value zero when all a_p are zero. To prevent this degeneration of the process we will introduce a further condition

$$\int_0^1 A(r, r_0)\,dr = 1. \tag{10.1.6}$$

Thus in averaging $\rho(r)$ by (10.1.3) we keep the total weight unity.

Let us introduce

$$g_p = \int_0^1 G_p(r)\,dr.$$

Then (10.1.6) becomes

$$\sum_1^P a_p g_p = 1, \tag{10.1.7}$$

and subject to this condition on $a_p(r_0)$ we have to minimise

$$S = 12\int_0^1 \sum_{p=1}^{P}\sum_{q=1}^{P} a_p a_q G_p G_q (r-r_0)^2\,\mathrm{d}r$$

$$= \sum_{p=1}^{P}\sum_{q=1}^{P} S_{pq} a_p a_q, \qquad (10.1.8)$$

where

$$S_{pq} = 12\int_0^1 G_p G_q (r-r_0)^2\,\mathrm{d}r \qquad (10.1.9)$$

is symmetric with respect to p and q.

If in the expression (10.1.5) for S we put

$$A(r, r_0) = \begin{cases} 1/2h \text{ in } r_0 - h \leq r \leq r_0 + h, \\ 0 \quad \text{otherwise,} \end{cases}$$

then $S = 2h$. This is the reason for the designation 'spread'. The value of S will be smaller (subject always to the constraint (10.1.6)) the closer the concentration of weighting around r_0. Thus S measures the *resolving power* of the linear combination $A(r, r_0)$ of data kernels.

We now utilise the summation convention, a duplicated suffix indicating summation over all values from 1 to P. We see that we are faced with a classical variational problem: to find the vector a_p which minimises the symmetric quadratic form

$$S_{pq} a_p a_q, \qquad [(10.1.8)]$$

subject to the condition

$$a_p g_p = 1. \qquad [(10.1.7)]$$

Geometrically, we have to find the shortest radius vector to a point on a hyper-ellipsoid $S_{pq} a_p a_q = \text{const}$ which lies also on the hyper-plane $a_p g_p = 1$. Algebraically, we employ a Lagrange-multiplier:

$$\delta(S_{pq} a_p a_q - 2\lambda a_p g_p) = 0,$$

that is,

$$(S_{pq} a_q - \lambda g_p)\delta a_p = 0 \qquad (10.1.10)$$

for arbitrary δa_p. Hence, if a_q^0 is the solution eigenvector corresponding to the eigenvalue λ^0,

$$S_{pq}a_q^0 - \lambda^0 g_p = 0. \qquad (10.1.11)$$

Thus

$$a_q^0 = \lambda^0 S_{pq}^{-1} g_p,$$

so that

$$1 = a_q^0 g_q = \lambda^0 S_{pq}^{-1} g_p g_q.$$

Hence

$$\lambda^0 = (S_{pq}^{-1} g_p g_q)^{-1}, \qquad (10.1.12)$$

and

$$a_q^0 = S_{pq}^{-1} g_p (S_{rs}^{-1} g_r g_s)^{-1}. \qquad (10.1.13)$$

It is easy to show that this vector gives a true minimum, and is the radius vector to the point at which one member of the family of ellipsoids touches the hyperplane. The corresponding spread is $S^0 = S_{pq}a_p^0 a_q^0 = S_{pq}a_p^0 \lambda^0 S_{rq}^{-1} g_r = \lambda^0$.

Returning to (10.1.4) we find the weighted average $\rho_a(r_0, m)$ at r_0 as

$$a_p^0 \gamma_p = S_{pq}^{-1} g_q \gamma_p (S_{rs}^{-1} g_r g_s)^{-1}, \qquad (10.1.14)$$

and for error-free observations $\bar{\gamma}_p$:

$$\rho_a(r_0, m_E) = S_{pq}^{-1} g_q \bar{\gamma}_p (S_{rs}^{-1} g_r g_s)^{-1}. \qquad (10.1.15)$$

This is the average of ρ which is spread least about r_0, and it is derived from the data $\bar{\gamma}_p$. All models which fit the data will give the same weighted average, since details of the model have disappeared from (10.1.15).

10.1.3 *Data subject to error*

So far we have assumed that the data $\bar{\gamma}_p$ constitute precise and accurate measures of the gross Earth functionals γ_p. But in fact our estimate of γ_p is normally the mean $\bar{\gamma}_p$ of a set of observations of which a typical one may be represented by $\bar{\gamma}_p + \Delta\gamma_p$, where $\Delta\gamma_p$ is not the true error, but the deviation from the mean. Then $\overline{\Delta\gamma_p} = 0$. Let the variance matrix of deviations be $E_{pq} \equiv \overline{\Delta\gamma_p \Delta\gamma_q}$, which is known directly from the observations. $\bar{\gamma}_p$ now represents not one precise observation, but the mean of a set which contains errors, and E_{pq} is an estimate of the variance of errors.

Let $\Delta\rho_a$ be the deviation in ρ_a consequent upon deviations $\Delta\gamma_p$ in γ_p. Then since in (10.1.15) ρ_a is a linear function of γ_p,

$$\bar{\rho}_a = a_q \bar{\gamma}_q, \tag{10.1.16}$$

and

$$\rho_a = \bar{\rho}_a + \Delta\rho_a = a_q \bar{\gamma}_q + a_q \Delta\gamma_q .$$

We define the variance of the average,

$$\begin{aligned} E &= \overline{\Delta\rho_a^2} \\ &= \overline{(a_p \Delta\gamma_p)(a_q \Delta\gamma_q)} \\ &= E_{pq} a_p a_q . \end{aligned} \tag{10.1.17}$$

When $a_p = a_p^0$, chosen uniquely to minimise S, E is determined and cannot be reduced. E is a measure of the lack of precision of our estimate ρ_a, arising because of lack of precision in the data, and it is obviously desirable to reduce it if possible by our choice of a_p. We can do so if we are prepared to relax the demand on S. Then we may hope to increase the precision of our estimate $\rho_a(r_0)$ – but at the expense of losing some sharpness of resolution.

So instead of minimising either S or E we will try to minimise $W = E + \sigma S$, still subject to the constraint on weighting: $a_p g_p = 1$. The resulting vector a_p will be a function of σ, and so will the corresponding spread $S(\sigma)$ and error $E(\sigma)$. So we make

$$\delta(W_{pq} a_p a_q - 2\mu a_p g_p) = 0, \tag{10.1.18}$$

where 2μ is a Lagrange-multiplier. This problem is formally identical with that solved when we minimised S, and the solution is

$$a'_p(\sigma) = W_{pq}^{-1}(\sigma) g_q \{W_{rs}^{-1}(\sigma) g_r g_s\}^{-1},$$

with corresponding

$$(E_{pq} + \sigma S_{pq}) a'_q = \mu g_p, \tag{10.1.19}$$

and

$$S(\sigma) = S_{pq} a'_p a'_q = s(\sigma), \tag{10.1.20}$$

$$E(\sigma) = E_{pq} a'_p a'_q = e(\sigma). \tag{10.1.21}$$

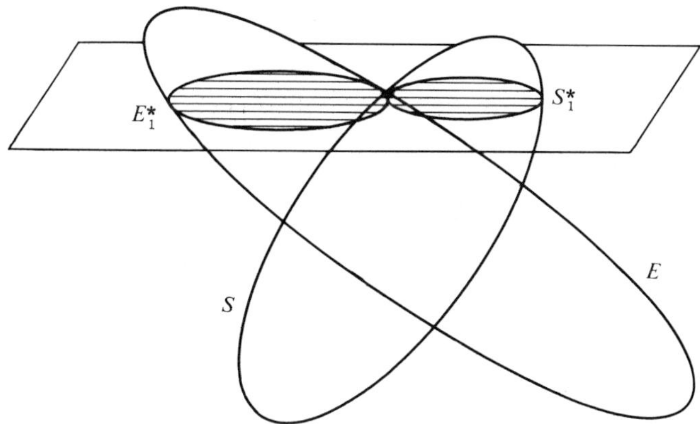

Fig. 10.1. Conics S_1^* and E_1^* seen as curves of intersection of members of families of hyper-ellipsoids $S(\sigma) = \text{const}$ and $E(\sigma) = \text{const}$ with hyperplane.

Geometrically, we see that the family of similar hyper-ellipsoids $S =$ const meets the hyperplane $a_p g_p = 1$ in a nest S^* of similar non-intersecting conics, and so does the family $E =$ const. We seek a point on the hyperplane and also on one of each family. If $S(\sigma)$ is given, a conic S_1^* of S^* is determined, and the smallest e will be that of the member of the family which meets the hyperplane in a conic E_1^* of E^* which touches S_1^*, as shown in Fig. 10.1. Since e and s are measures of the size of E and S, it appears from the figure that e decreases as s increases.

We now investigate the relation between $s(\sigma)$ and $e(\sigma)$. Let us denote derivatives with respect to σ by a dot. Then

$$g_p \dot{a}'_p = 0,$$
$$2S_{pq} \dot{a}'_p a'_q = \dot{s},$$
$$2E_{pq} \dot{a}'_p a'_q = \dot{e},$$

and from (10.1.19)

$$E_{pq} \dot{a}'_p a'_q + \sigma S_{pq} \dot{a}'_p a'_q = \mu g_p \dot{a}'_p = 0.$$

Hence

$$\dot{e} + \sigma \dot{s} = 0 \qquad \text{and} \qquad \mathrm{d}e/\mathrm{d}s = -\sigma. \tag{10.1.22}$$

Since we know that $\mathrm{d}e/\mathrm{d}s$ is negative, we need consider only positive

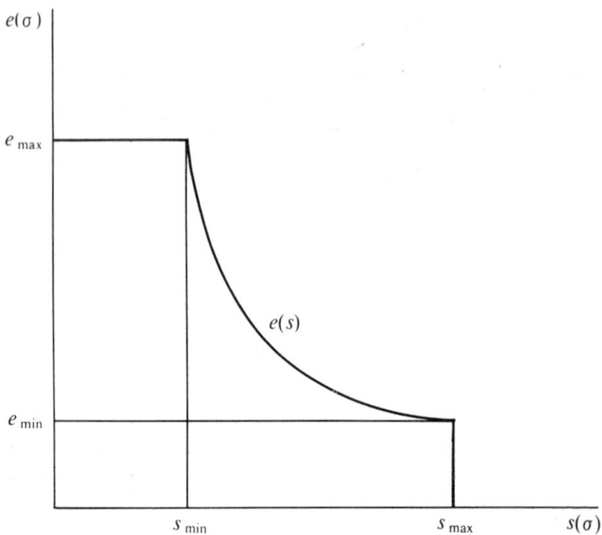

Fig. 10.2. Trade-off curve $e(s)$.

values of σ. Again from (10.1.19)

$$e+\sigma s=\mu,$$

and so

$$\dot{\mu}=s. \tag{10.1.23}$$

When σ becomes infinite we return to our earlier problem, with solution a_q^0, $s=s_{\min}$, $e=e_{\max}$. When $\sigma=0$ we have the same problem with E replacing S, and $s=s_{\max}$, $e=e_{\min}$. It can be shown (Backus & Gilbert, 1970) that as σ proceeds from zero to infinity, e increases monotonically from $e_{\min}$ to $e_{\max}$ and s falls monotonically from $s_{\max}$ to $s_{\min}$. The curve graphing $e(\sigma)$ against $s(\sigma)$ is therefore as shown in Fig. 10.2. It can be shown that the slope of $e(s)$ is zero at $s_{\max}$ and infinite at $s_{\min}$. Consequently an increase of s above $s_{\min}$ gives a large decrease in $e(s)$, and an increase of e above $e_{\min}$ gives a large decrease in s. The name 'trade-off curve' neatly sums up the character and use of the curve.

The following example of the employment of data-kernels has been given by Gilbert (1971). It is required to determine averaging kernels for $\rho(r)$ and the corresponding local averages $\rho_a(r_0)$, given the gross Earth functionals:

Mass, Moment of inertia about axis,

$$\left.\begin{array}{l}\text{Modes: } {}_0S_0,\ {}_1S_0,\ {}_2S_0,\ {}_3S_0,\ {}_1S_1,\ {}_2S_1,\\ {}_0S_2,\ {}_2S_2,\ {}_1S_3,\ {}_0S_4,\ {}_1S_4,\ {}_2S_4,\\ {}_4S_4,\ {}_0S_7,\ {}_1S_8,\ {}_0S_{25},\ {}_0S_{49},\ {}_0S_{73},\ {}_0S_{97},\\ {}_0T_7,\ {}_0T_{14},\ {}_0T_{27},\ {}_0T_{53},\ {}_0T_{105}.\end{array}\right\} \qquad (10.1.24)$$

We assume that $\lambda(r)$ and $\mu(r)$ are known exactly, and that the data are free from error: thus the problem of trade-off between precision and resolution does not arise. Corresponding to the 26 gross Earth functionals there are 26 data-kernels G_p: from these we construct, for each point r_0 at which an average is required, a linear combination

$$A(r, r_0) = \sum_{1}^{26} a_p(r_0) G_p$$

We minimise[†]

$$K = \int_0^1 \{A(r, r_0) - \delta(r - r_0)\}^2 (r - r_0)^2 \,\mathrm{d}r, \qquad (10.1.25)$$

subject to $\int_0^1 A\,\mathrm{d}r = 1$. The appropriate choice of $a_p(r_0)$ for each r_0 determines the averaging kernel $A(r, r_0)$ for that point. These kernels are shown in Fig. 10.3 for 12 values r_0. We see from the shape of the peaks that the selected set of gross Earth functionals (10.1.24) gives sharp resolution in the neighbourhood of the core–mantle boundary and in the upper mantle, but very weak resolution in the core and lower mantle.

Local average densities may now be computed. They are shown for 35 points in Fig. 10.4. For each point, the averaging kernel employed is that at the nearest r_0.

For an example of the use of the trade-off diagram when the errors in the data are taken into consideration, the reader may refer to the original paper of Backus & Gilbert (1970), or to Gilbert (1971).

Thus the Backus–Gilbert technique gives a way of estimating the precision and resolution which can be derived from a given data-set, provided the problem is linear or can be linearised. For further discussion of the inversion problem refer to Parker (1977).

[†] The use of K in place of (10.1.5) is explained by Gilbert (1971).

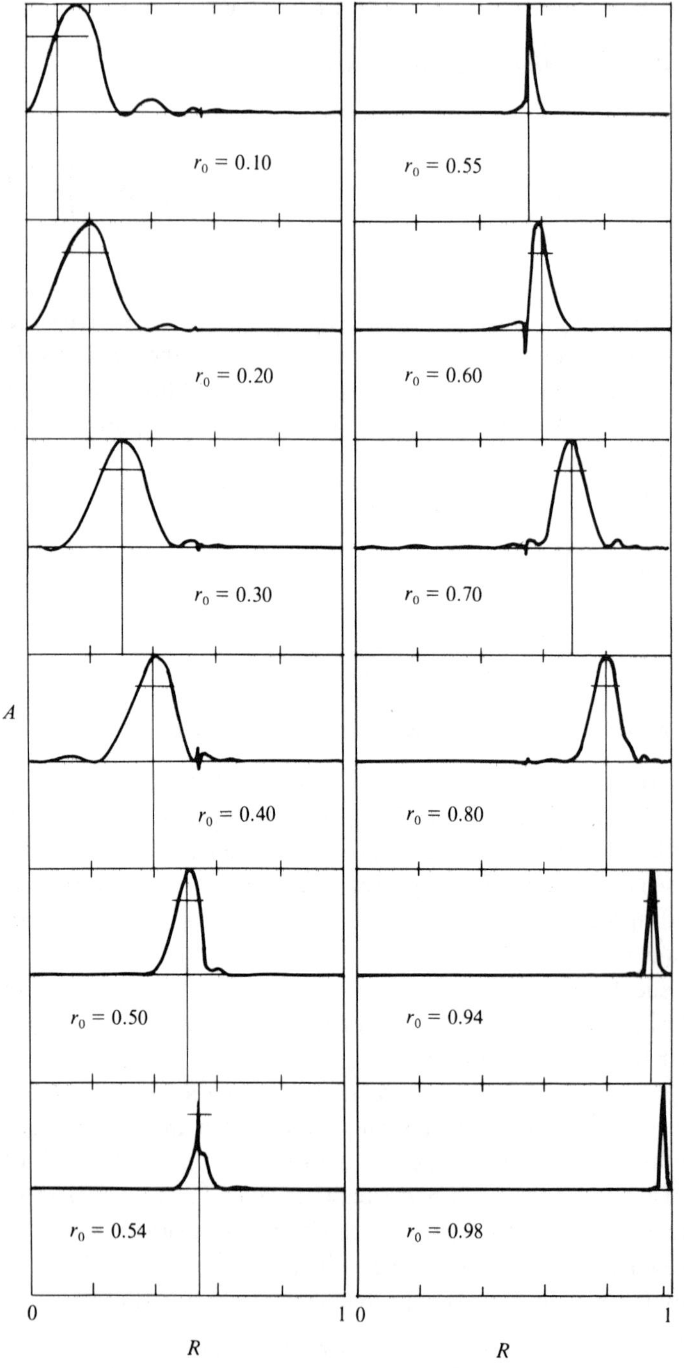

Fig. 10.3. Averaging kernels $A(r, r_0)$ for ρ (using K criterion). r_0 is located by ordinate and spread s by horizontal bar. (From Gilbert, 1971.)

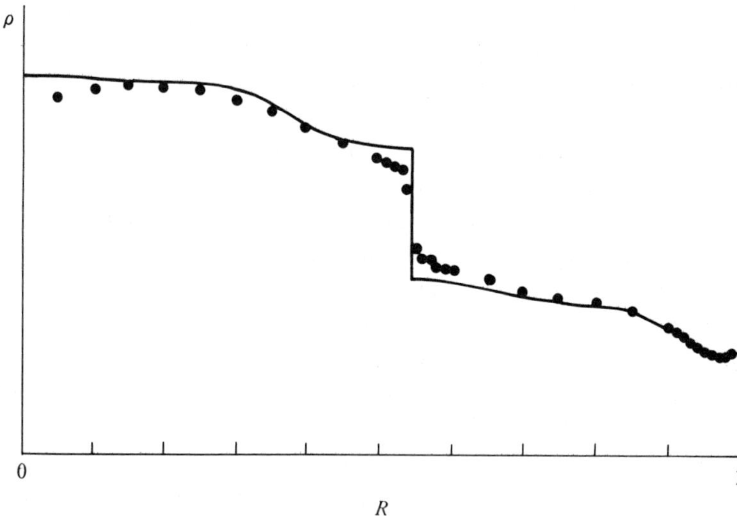

Fig. 10.4. Dots show local average density computed with kernels shown in Fig. 10.3. Solid line shows the starting model density. (From Gilbert, 1971.)

Direct non-linear inversion, which is ideally required for the construction and testing of Earth-models derived from free oscillation data, is not yet practicable, but progress has been made in inversion for asymptotic distribution of both toroidal and spheroidal eigenfrequencies (Kennett & Woodhouse, 1978, Brodskii & Levshin, 1979).

10.2 Allowance for lateral inhomogeneity, anisotropy and pre-stress

In Chapter 2, where we discussed the eigenfrequencies of a spherically symmetric, non-rotating, isotropic elastic sphere, we pointed out that a mode of the type $W_n(r)P_l^m(\cos\vartheta)\exp(im\varphi)$ $\exp i\,{}_n\omega_l t$ has frequency ${}_n\omega_l$ independent of m because axial symmetry of the model and the absence of rotation cause the multiplet of $2l+1$ frequencies to degenerate to single value. In Chapter 7, we showed that the spin of the Earth removes that degeneracy, and spreads the frequencies of the multiplet. We stated, but did not prove, that other departures from symmetry have the same effect.

In this section we show how to obtain the modifications of eigenfrequencies due to lateral inhomogeneities, anisotropy and initial stress. (Evidence of anisotropy in the mantle is given by Crampin (1977) and Bamford & Crampin (1977).) We group these together because each of them produces small modifications, which can be found by perturbation methods. The perturbation analysis is essentially the same for all; we follow the inclusive treatment by Woodhouse & Dahlen (1978).

We envisage our new Earth-model as a quasi-spherical body with internal layering such that no two interfaces intersect anywhere. The union of interior and exterior boundaries we call Σ, and the remainder of space, not occupied by Σ, we call V. We intend to compare eigenfrequencies of a reference model and a perturbed model, that is, a model where certain physical parameters have been slightly changed.

We take for the reference model one which is initially at rest with respect to a set of rectangular Cartesian axes Ox_i, which are rotating with uniform angular velocity $\mathbf{\Omega}$ about the centre of mass O. Let $\mathbf{x}$ be any point of Σ or V, and $\hat{\mathbf{n}}$ the unit outward normal at any point on Σ. In the initial (non-vibrating) state of the reference model, let $\rho_0, \psi_0, \mathbf{T}_0$ be the density, gravitational potential, and static stress tensor. The centripetal potential will be denoted by χ, where

$$\chi = \tfrac{1}{2}\{\Omega^2 r^2 - (\Omega_k x_k)^2\}. \tag{10.2.1}$$

The field equations in this state are

$$\rho_0 \nabla(\psi_0 + \chi) = -\nabla \cdot \mathbf{T}_0, \tag{10.2.2}$$

$$\nabla^2 \psi_0 = -4\pi G \rho_0. \tag{10.2.3}$$

The boundary conditions on Σ are

$$[\psi_0]_-^+ = 0, \tag{10.2.4}$$

$$[\hat{\mathbf{n}} \cdot \nabla \psi_0]_-^+ = 0, \tag{10.2.5}$$

$$[\hat{\mathbf{n}} \cdot \mathbf{T}_0]_-^+ = 0. \tag{10.2.6}$$

The third of these boundary conditions states that the stress vector on an element of Σ is continuous across Σ. In the liquid part of the core the stress is isotropic, so that if the normal component $\hat{\mathbf{n}} \cdot \mathbf{T}_0 \cdot \hat{\mathbf{n}}$ of the stress vector on an element of Σ is denoted by π_0, then at a fluid–solid boundary $\hat{\mathbf{n}} \cdot \mathbf{T}_0 = \pi_0 \hat{\mathbf{n}}$.

Now consider a small free vibration of our reference model. In a normal mode of angular frequency ω, let the associated displacement eigenfunction be $\mathbf{u}$, and let the corresponding increments in density and gravitational potential be ρ_1, ψ_1, where

$$\rho_1 = -\nabla\cdot(\rho_0\mathbf{u}). \tag{10.2.7}$$

Then the field equations of the vibration are

$$\nabla^2\psi_1 = -4\pi G\rho_1, \tag{10.2.8}$$

$$\rho_0\{-\omega^2\mathbf{u} + 2i\omega\boldsymbol{\Omega}\wedge\mathbf{u} - \nabla\psi_1 - \mathbf{u}\cdot\nabla\nabla(\psi_0+\chi)\} = \nabla\cdot\tilde{\mathbf{T}}, \tag{10.2.9}$$

where the tensor $\tilde{\mathbf{T}}$ is the increment, at the material point, of the Piola–Kirchhoff stress (Malvern, 1969, Dahlen, 1972). It is given in terms of the local strain e_{ij}, local rotation ω_{ij}, and the initial stress ($\mathbf{T}_0$, with components T^0_{ij}) by

$$\tilde{T}_{ij} = c_{ijkl}e_{kl} + (\omega_{ik}T^0_{kj} - T^0_{ik}\omega_{kj}) + \tfrac{1}{2}e_{kk}T^0_{ij} + \tfrac{1}{2}T^0_{kl}e_{kl}\delta_{ij} - T^0_{jk}u_{i,k}. \tag{10.2.10}$$

We write this as $\tilde{T}_{ij} = \Lambda_{ijkl}u_{k,l}$, thus defining the tensor $\boldsymbol{\Lambda}$. The term $c_{ijkl}e_{kl}$ is the usual expression for that part of the stress which is derived from the local strain by Hooke's Law, c_{ijkl} being the elastic constants in a general anisotropic medium, with symmetries

$$c_{ijkl} = c_{jikl} = c_{klij} = c_{ijlk}. \tag{10.2.11}$$

The terms containing $\boldsymbol{\omega}$ appear because the original stress field $\mathbf{T}_0$ is changed by local rotation from T^0_{ij} to $(\delta_{ki}+\omega_{ki})T^0_{kl}(\delta_{lj}+\omega_{lj})$, and we keep only first-order terms in the increment. The other three terms on the right-hand side of (10.2.10) arise from the interplay of strain and original stress, and are found by satisfying the condition that the energy in the final state is independent of the path by which that state is reached.

In constructing boundary conditions, we make use of the *surface gradient*

$$\nabla_\Sigma \equiv \nabla - \hat{\mathbf{n}}(\hat{\mathbf{n}}\cdot\nabla) \tag{10.2.12}$$

at any point of Σ, and allow for the shifting of interfaces in the disturbance, so that on Σ, instead of $\tilde{\mathbf{T}}$ we deal with

$$\mathbf{t} = \hat{\mathbf{n}}\cdot\tilde{\mathbf{T}} - \hat{\mathbf{n}}\nabla_\Sigma\cdot(\pi_0\mathbf{u}) + \pi_0(\nabla_\Sigma\mathbf{u})\cdot\hat{\mathbf{n}} \tag{10.2.13}$$

at the undeformed boundary. Instead of $\hat{\mathbf{n}}\cdot\nabla\psi_1$ we deal with

$$g_1 = \hat{\mathbf{n}}\cdot\left(-\frac{1}{4\pi G}\nabla\psi_1 + \rho_0\mathbf{u}\right) \tag{10.2.14}$$

at the undeformed boundary. Then the boundary conditions can be set out as follows:

$$\begin{aligned}
&\text{welded boundaries:} && [\mathbf{u}]_-^+ = 0, \quad [\mathbf{t}]_-^+ = 0,\\
&\text{fluid–solid boundaries:} && [\hat{\mathbf{n}}\cdot\mathbf{u}]_-^+ = 0, \quad [\mathbf{t}]_-^+ = 0 \text{ (here } \mathbf{t} = \hat{\mathbf{n}}(\hat{\mathbf{n}}\cdot\mathbf{t})\text{)},
\end{aligned} \tag{10.2.15}$$

$$\begin{aligned}
&\text{free boundaries:} && \mathbf{t} = 0,\\
&\text{all boundaries:} && [\psi_1]_-^+ = 0, \quad [g_1]_-^+ = 0.
\end{aligned} \tag{10.2.16}$$

We will suppose that eigenvalues and corresponding eigenfunctions have been calculated for the reference model. Usually the reference model is taken to be spherically symmetric, isotropic, and with initial static stress which is purely hydrostatic: $T_{ij}^0 = -p_0\delta_{ij}$. It may also be non-rotating, so that we can use the methods of Chapters 5 and 6. If the reference model is rotating, the effect of rotation is included as in Chapter 7.

Now we perturb the model, so that physical parameters which in the reference model are $\rho_0, \Psi_0 \equiv \psi_0 + \chi, \mathbf{T}_0$ and $\mathbf{\Lambda}$ take small increments $\delta\rho_0, \delta\Psi_0, \delta\mathbf{T}_0, \delta\mathbf{\Lambda}$ at points fixed in a reference frame rotating with perturbed angular velocity $\mathbf{\Omega} + \delta\mathbf{\Omega}$. It is possible to include also perturbation of the boundary Σ (Woodhouse & Dahlen, 1978), but here we leave the boundary unchanged. When this perturbed model vibrates, let the perturbed eigenfrequency be $\omega + \delta\omega$. We assume that all perturbations are first-order small quantities, and neglect second-order terms.

Then either by use of Rayleigh's Principle or directly from the perturbed field equations and boundary conditions we can obtain (Woodhouse & Dahlen, 1978) a formula for $\delta\omega$, which is the increment in angular frequency of a particular mode of free oscillation:

$$\delta\omega\int_V (2\omega\rho_0 u_i u_i^* - 2\mathrm{i}\rho_0\varepsilon_{ijk}\Omega_i u_j u_k^*)\mathrm{d}V$$

$$= \int_V \{2\mathrm{i}\omega\rho_0\varepsilon_{ijk}\delta\Omega_i u_j u_k^* + \delta\rho_0(-\omega^2 u_i u_i^* + 2\mathrm{i}\omega\varepsilon_{ijk}\Omega_i u_j u_k^* -$$

$$- u_i u_j^* \Psi_{0,ij} - u_i \psi_{1,i}^* - u_i^* \psi_{1,i}) - \rho_0 u_i u_j^* \delta\Psi_{0,ij} + \delta\Lambda_{ijkl} u_{j,i} u_{l,k}^* \} \mathrm{d}V. \tag{10.2.17}$$

This formula gives $\delta\omega$ in terms of the *unperturbed* eigenfunctions $\mathbf{u}, \psi_1$, their complex conjugates $\mathbf{u}^*, \psi_1^*$, and the perturbations $\delta\boldsymbol{\Omega}, \delta\rho_0, \delta\Psi_0, \delta\boldsymbol{\Lambda}$ of the Earth-model, where $\delta\boldsymbol{\Lambda}$ includes effects of anisotropy and pre-stress, and $\delta\boldsymbol{\Omega}, \delta\rho_0$ may be independently assigned. $\delta\Psi_0$, on the other hand, must be evaluated from the condition of equilibrium of the perturbed model in its initial state. Lateral inhomogeneities, such as the differences between oceanic and continental crust and mantle, will contribute to $\delta\rho_0$ and $\delta\boldsymbol{\Lambda}$, which may be functions of r, ϑ and φ.

In (10.2.17) the three terms containing $\boldsymbol{\Omega}$ or $\delta\boldsymbol{\Omega}$ are due to Coriolis forces, those containing $\delta\rho_0$ to density changes. The term with $\delta\Psi_{0,ij}$ represents a change in gravitational and centripetal energy due to changed spin, while the term with $\omega\delta\omega$ represents the change in kinetic energy of vibration due to the change in eigenfrequency of the mode. The term containing $\delta\boldsymbol{\Lambda}$ comes from the change in strain energy due to pre-stress and the modification of elastic constants.

The application of formula (10.2.17) is described in detail by Woodhouse & Dahlen (1978); lateral heterogeneity has also been treated by Zharkov & Lyubimov (1970a, b), Madariaga (1972) and Luh (1973, 1974).

10.3 Anelasticity

10.3.1 *Generalised linear anelastic solids*

We start our investigation of the departure from perfect elasticity (following Kanamori & Anderson (1977) and Woodhouse (1977)) by generalising the one-dimensional stress–strain relationship

$$\sigma = M\varepsilon, \tag{10.3.1}$$

where ε is strain and σ is stress. M is the elastic constant, which may be, for example, Young's modulus or the rigidity. We generalise (10.3.1) to

$$\sigma(t) + \tau_\sigma \dot{\sigma}(t) = M_{\mathrm{R}} \{\varepsilon(t) + \tau_\varepsilon \dot{\varepsilon}(t)\}, \tag{10.3.2}$$

where τ_σ and τ_ε are constants with the dimension of time, and M_{R}

has the same dimensions as M. This formula describes the behaviour of a *standard linear* solid; the Kelvin–Voigt Law is obtained as a special case by putting $\tau_\sigma = 0$.

We may proceed neatly to the integrated form of (10.3.2) by use of the Laplace transform. Suppose that a stress $\sigma(t)H(t)$ is applied. Then with the usual notation (p being the transform variable and $\bar{\sigma}(p)$ the transform of $\sigma(t)$)

$$\bar{\sigma} + \{p\bar{\sigma} - \sigma(0)\}\tau_\sigma = M_{\mathrm{R}}[\bar{\varepsilon} + \{p\bar{\varepsilon} - \varepsilon(0)\}\tau_\varepsilon],$$

so that

$$\bar{\varepsilon}(p) = \frac{1}{M_{\mathrm{R}}(1 + p\tau_\varepsilon)}\{(1 + p\tau_\sigma)\bar{\sigma} - \tau_\sigma\sigma(0) + M_{\mathrm{R}}\tau_\varepsilon\varepsilon(0)\}.$$

Inverting the transform by means of the convolution theorem we get

$$\varepsilon(t) = \int_0^t \dot{\sigma}(t')\phi(t - t')\mathrm{d}t' + \sigma(0)\phi(t) + \left\{\varepsilon(0) - \frac{1}{M_{\mathrm{R}}}\frac{\tau_\sigma}{\tau_\varepsilon}\sigma(0)\right\}\mathrm{e}^{-t/\tau_\varepsilon} \tag{10.3.3}$$

where

$$\phi(t) = \frac{1}{M_{\mathrm{R}}}\{1 - (1 - \tau_\sigma/\tau_\varepsilon)\mathrm{e}^{-t/\tau_\varepsilon}\}. \tag{10.3.4}$$

The terms in (10.3.3) in the curled bracket represent transients depending on $\varepsilon(0)$ and $\sigma(0)$. These and the form of $\phi(t)$ in (10.3.4) show that τ_ε is the time required for the amplitude of $\exp(-t/\tau_\varepsilon)$ to fall to $\exp(-1)$ of its original value. τ_ε is therefore the *relaxation time*

If $\sigma(t) = H(t - T)$, so that unit stress is suddenly applied at time $T(>0)$ and then maintained,

$$\dot{\sigma}(t) = \delta(t - T),$$

and

$$\varepsilon(t) = \phi(t - T)H(t - T). \tag{10.3.5}$$

Thus $\phi(t - T)$ is the response to a step-function of stress applied at $t = T$. From (10.3.5) we see that the initial strain is $\varepsilon(T) = \phi(0) = \tau_\sigma/(M_{\mathrm{R}}\tau_\varepsilon)$ (from (10.3.4)). Thus initially the stress is M_{U} times the strain, where

$$M_{\mathrm{U}} = M_{\mathrm{R}}\tau_\varepsilon/\tau_\sigma. \tag{10.3.6}$$

After a long time, $\phi(t) \to 1/M_{\mathrm{R}}$ as $t \to \infty$; that is, $\varepsilon(\infty) = \phi(\infty) = 1/M_{\mathrm{R}}$. So since $\sigma(T) = M_{\mathrm{U}}\varepsilon(T)$, $\sigma(\infty) = M_{\mathrm{R}}\varepsilon(\infty)$, M_{U} is called the *unrelaxed modulus* and M_{R} the *relaxed modulus*.

If in (10.3.3) we disregard transients and allow $\phi(t)$ to be a more general *creep function* which may come into operation before $t = 0$, we obtain Boltzmann's after-effect equation:

$$\varepsilon(t) = \int_{-\infty}^{t} \dot{\sigma}(t')\phi(t - t')\mathrm{d}t'.$$

$\phi(t - t')$ is a function of the lapse of time between t' (variable of integration) and t (when $\varepsilon(t)$ is evaluated); it serves therefore the function of a 'memory' of previous states, and is determined by anelastic properties of the medium.

The expression (10.3.4) for $\phi(t)$ pertains to a single mechanism of attenuation. But it is common for materials composed of many constituents to have several relaxation times which belong to different mechanisms which come into operation at different frequencies – such as interstitial atom relaxation, grain boundary relaxation, and phase changes. Then we may replace (10.3.4) by

$$\phi(t) = \frac{1}{M_R}\left\{1 - \sum\left(1 - \frac{\tau_\sigma}{\tau_\varepsilon}\right)\mathrm{e}^{-t/\tau_\varepsilon}\right\}, \tag{10.3.7}$$

Σ meaning a summation over contributions from different mechanisms. Then

$$\begin{aligned}\varepsilon(t) &= \int_{-\infty}^{t} \frac{\dot{\sigma}(t')}{M_R}\left\{1 - \sum\left(1 - \frac{\tau_\sigma}{\tau_\varepsilon}\right)\mathrm{e}^{-(t-t')/\tau_\varepsilon}\right\}\mathrm{d}t' \\ &= \frac{\sigma(t)}{M_R} - \frac{1}{M_R}\sum\left(1 - \frac{\tau_\sigma}{\tau_\varepsilon}\right)\int_0^\infty \mathrm{e}^{-\theta/\tau_\varepsilon}\dot{\sigma}(t - \theta)\mathrm{d}\theta. \end{aligned} \tag{10.3.8}$$

We are concerned with harmonic time-variation, so we replace $\dot{\sigma}(t - \theta)$ by $\mathrm{i}\omega\sigma(t - \theta)$ in the integrand of (10.3.8) and get

$$\begin{aligned}\varepsilon(t) &= \frac{1}{M_R}\sigma(t)\left\{1 - \sum\left(1 - \frac{\tau_\sigma}{\tau_\varepsilon}\right)\int_0^\infty \mathrm{i}\omega\mathrm{e}^{-\mathrm{i}\omega\theta}\mathrm{e}^{-\theta/\tau_\varepsilon}\mathrm{d}\theta\right\} \\ &= \frac{1}{M_R}\sigma(t)(1 - A - \mathrm{i}B), \end{aligned} \tag{10.3.9}$$

where

$$\left.\begin{aligned} A(\omega) &= \sum\frac{\omega^2\tau_\varepsilon(\tau_\varepsilon - \tau_\sigma)}{1 + \omega^2\tau_\varepsilon^2}, \\ B(\omega) &= \sum\frac{\omega(\tau_\varepsilon - \tau_\sigma)}{1 + \omega^2\tau_\varepsilon^2}. \end{aligned}\right\} \tag{10.3.10}$$

It can be shown (Liu, Anderson & Kanamori, 1976) that if W is the elastic strain energy per unit cycle per unit volume, and ΔW is the energy dissipated per unit cycle per unit volume, then $\Delta W/W = 2\pi B$. Experiment (and common observation) shows that this is positive for any real mechanism, so that $\tau_\varepsilon > \tau_\sigma$ and, by (10.3.6), $M_{\mathrm{U}} > M_{\mathrm{R}}$. For a simple but informative analysis we assume that τ_σ and τ_ε are linearly related, and write $\tau_\varepsilon = \tau$,

$$\tau_\sigma = (1 - C)\tau, \tag{10.3.11}$$

where C is a small constant which is positive since $\tau_\sigma < \tau_\varepsilon$. Then if the summations over relaxation mechanisms in (10.3.10) are replaced by integrations of a distribution $D(\tau)$,

$$A(\omega) = \int_\tau \frac{C\omega^2\tau^2}{1 + \omega^2\tau^2} D(\tau)\,\mathrm{d}\tau, \tag{10.3.12}$$

$$B(\omega) = \int_\tau \frac{C\omega\tau}{1 + \omega^2\tau^2} D(\tau)\,\mathrm{d}\tau. \tag{10.3.13}$$

We have no direct knowledge of $D(\tau)$ within the Earth, but we have a fairly well-established fact that the attenuation of travelling waves or eigenvibrations is almost independent of frequency over the range of seismic frequencies. We find that we can produce such an independence by the assumption that $D(\tau)$ has the simple form

$$\left.\begin{aligned} D(\tau) &= 1/\tau \qquad \text{when } \tau_1 \le \tau \le \tau_2, \\ &= 0 \qquad \text{otherwise.} \end{aligned}\right\} \tag{10.3.14}$$

Then

$$A(\omega) = \frac{1}{2} C \ln \frac{1 + \omega^2\tau_2^2}{1 + \omega^2\tau_1^2}, \tag{10.3.15}$$

$$B(\omega) = C \tan^{-1} \frac{\omega(\tau_2 - \tau_1)}{1 + \omega^2\tau_1\tau_2}. \tag{10.3.16}$$

We can now find the modification of a mode of vibration due to anelasticity. Returning to equation (10.3.1)

$$\sigma = M\varepsilon,$$

which related stress and strain in a perfectly elastic medium, we can

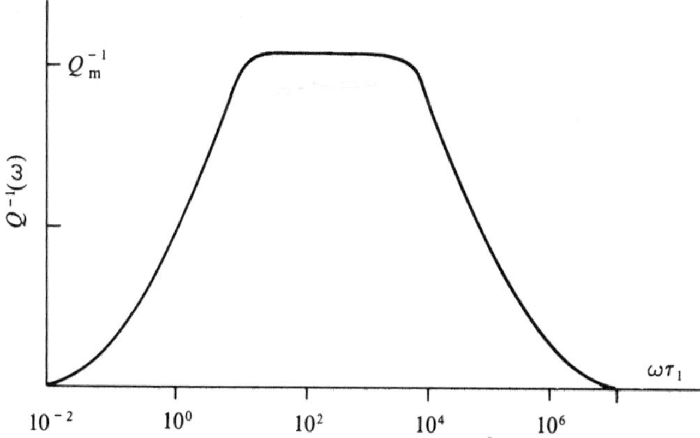

Fig. 10.5. Constant Q model derived from linear visco-elastic model. The curve was computed for $\tau_2/\tau_1 = 10^5$. (After Kanamori & Anderson, 1977.)

set down an equation of motion for vibrations:

$$K\ddot{\varepsilon} = -\sigma = -M\varepsilon,$$

where K is a constant. For vibrations of angular frequency ω,

$$\omega^2 = M/K. \tag{10.3.17}$$

For small departures from perfect elasticity, A and B in (10.3.9) are small, so that to the first order of small quantities we have $\sigma = M(1 + A + \mathrm{i}B)\varepsilon$,
and

$$(\omega + \delta\omega)^2 = M(1 + A + \mathrm{i}B)/K,$$

giving

$$\delta\omega/\omega = \tfrac{1}{2}(A + \mathrm{i}B), \tag{10.3.18}$$

approximately. Thus the modified vibration is given by

$$\exp \mathrm{i}(1 + \tfrac{1}{2}A + \tfrac{1}{2}\mathrm{i}B)\omega t = \exp(-\tfrac{1}{2}B\omega t) \exp \mathrm{i}(1 + \tfrac{1}{2}A)\omega t.$$

The factor $\exp(-\tfrac{1}{2}B\omega t)$, which measures attenuation, is usually expressed in terms of a *quality factor* $Q = 1/B$. Fig. 10.5 shows Q^{-1} graphed against ω; it shows that for a wide central range of frequencies Q^{-1} is almost constant. For very large and very small values of ω the value of Q^{-1} drops away, that is, the quality factor becomes large. These are regions of ω far from those where the relaxation mechanisms are most efficient.

From (10.3.16) we see that the greatest value of attenuation, which we will denote by $Q_m^{-1} \equiv [Q^{-1}]_{\max}$, is $\frac{1}{2}C\pi$. Thus

$$C = 2/(\pi Q_{\mathrm{m}}),$$

and

$$Q^{-1} = \frac{2}{\pi} Q_{\mathrm{m}}^{-1} \tan^{-1} \frac{\omega(\tau_2 - \tau_1)}{1 + \omega^2 \tau_1 \tau_2}. \tag{10.3.19}$$

10.3.2 *Modification of eigenfrequencies*

Turning now to oscillations of the Earth, we see that M may represent either $\lambda + 2\mu$ (P waves, Rayleigh waves, spheroidal vibrations) or μ (S waves, Love waves, toroidal vibrations). To illustrate our method, we deal with the latter. The velocity of an S wave of angular frequency ω will be

$$\begin{aligned} c_{\mathrm{S}}(\omega) &= (M_{\mathrm{R}}/\rho)^{1/2}\left(1 + \frac{1}{2} C \ln \frac{1 + \omega^2 \tau_2^2}{1 + \omega^2 \tau_1^2}\right)^{1/2} \\ &\approx (M_{\mathrm{R}}/\rho)^{1/2}\left(1 + \frac{1}{4} C \ln \frac{1 + \omega^2 \tau_2^2}{1 + \omega^2 \tau_1^2}\right), \end{aligned} \tag{10.3.20}$$

when C is small. Thus, comparing waves of angular frequencies ω_1 and ω_2 we have

$$\begin{aligned} \frac{c_{\mathrm{S}}(\omega_1)}{c_{\mathrm{S}}(\omega_2)} &\approx \left(1 + \frac{1}{4} C \ln \frac{1 + \omega_1^2 \tau_2^2}{1 + \omega_1^2 \tau_1^2}\right)\left(1 - \frac{1}{4} C \ln \frac{1 + \omega_2^2 \tau_2^2}{1 + \omega_2^2 \tau_1^2}\right) \\ &\approx 1 + \frac{1}{4} C \ln \left(\frac{1 + \omega_1^2 \tau_2^2}{1 + \omega_1^2 \tau_1^2} \frac{1 + \omega_2^2 \tau_1^2}{1 + \omega_2^2 \tau_2^2}\right). \end{aligned} \tag{10.3.21}$$

Now let $1/\tau_2 \ll \omega_1 < \omega_2 \ll 1/\tau_1$. Then

$$\begin{aligned} \frac{c_{\mathrm{S}}(\omega_1)}{c_{\mathrm{S}}(\omega_2)} &\approx 1 + \frac{1}{2} C \ln \frac{\omega_1}{\omega_2} \\ &= 1 + \frac{1}{\pi Q_{\mathrm{m}}} \ln \frac{\omega_1}{\omega_2}. \end{aligned} \tag{10.3.22}$$

This expression does not depend on knowledge of τ_1 and τ_2, and

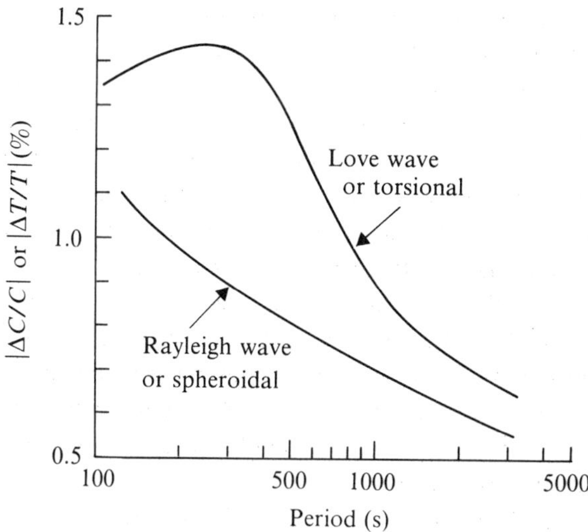

Fig. 10.6. Fractional change in Love wave (toroidal mode) and Rayleigh wave (spheroidal mode) phase-velocities (periods) as functions of period, computed from observed Q. (From Kanamori & Anderson, 1977.)

gives the correction needed to derive c_S at frequency ω_1, relative to $c_S(\omega_2)$, within the range where $Q \approx Q_m$.

It is shown by Liu *et al.* (1976) how to derive from this the relative correction to the phase-velocity of Love waves, and consequently the relative corrections to the eigenfrequencies of toroidal oscillations by use of the formula $\omega = (l + \frac{1}{2})c_L/a$.

If a velocity distribution of c_S with depth has been constructed from body-wave data, with frequencies about 1 Hz, it is found that corrections of the order of 1% should be made before comparing computed toroidal frequencies or Love-wave dispersion curves with data obtained by observation of long surface waves and free oscillations. Fig. 10.6 shows the magnitude of the correction (which is in fact negative since $\omega_1 < \omega_2$) as a function of ω. It shows also the corrections for Rayleigh waves or spheroidal oscillations.

10.4 A Reference Earth-model

In 1940, the International Association for Seismology adopted the Jeffreys–Bullen tables of travel-times of earthquake phases for use in preparing the International Seismological Summary. These tables provided a widely known and understood reference, and it became common for seismologists to describe, for instance, a particular region, or a particular hypothesis as to Earth structure at depth, by stating how travel-times for that region or structure deviated from the Jeffreys–Bullen times. But these travel-times, while corresponding to Jeffreys's 1939 distributions of $c_P(r)$ and $c_S(r)$, did not determine the three parameters λ, μ and ρ separately as functions of depth. Such distributions were needed in computation of surface-wave dispersion curves and eigenfrequencies of toroidal and spheroidal oscillations of any proposed model Earth.

There existed also travel-time tables of Gutenberg and (more recently) Herrin *et al.*, with corresponding distributions of $c_P(r)$ and $c_S(r)$; density distributions $\rho(r)$ were proposed by Bullard and by Bullen (several different models). Confusion arose when different combinations of these distributions were used as reference models by later authors who wished to describe anomalous data or regional differences, or to propose new internal discontinuities or revised depths of a discontinuity – especially that at the core–mantle boundary. Some authors combined velocity-models and density-models which were in fact incompatible in having internal boundaries at different levels. Consequently it was extremely difficult to compare new proposals. If Professor A described his new model by reference to a 'Gutenberg–Bullen A′' Earth-model, while Dr. B described his in terms of a 'Jeffreys–Bullard' Earth-model, how could one discern without undue labour the essential points of difference of Professor A's and Dr. B's models?

In order to dispel such obscurity, and to make comparison of new models a straightforward matter, it was obviously desirable that all new models should be described in relation to a single recognised reference model. The reference model should be a good one, but it was not necessary to demand perfection.

So in 1971 the International Union of Geodesy and Geophysics set up the Standard Earth Model Committee, under the chairmanship of Professor K. E. Bullen. The Committee found considerable opposition to the idea of a single reference model. (The original term 'standard' was an unfortunate choice, as it was often taken to imply that the chosen model would have to be better than all others.)

The task of obtaining agreement on one model looked forbiddingly like that of getting a bunch of prima donnas to sing in unison (Bullen, 1974).

But progress was made, and in 1976 the Committee was able to invite seismologists to submit proposals for a reference Earth-model, which should be presented in terms of c_P, c_S and ρ specified as functions of position. The Committee did not impose constraints beyond those given in the following guidelines.

'We hope that any Earth model proposed would satisfy the following criteria:

(1) The model should be consistent with the following values, which are derived from constants adopted by the International Association of Geodesy:

  (a) Mean radius of the Earth $R = 6\,371.0$ km (as defined by the radius of the sphere of the same volume).
  (b) Flattening: 298.25.
  (c) Mass $M = 5.974 \times 10^{24}$ kg.
  (d) Ratio $I/MR^2 = 0.3308$ (where I is mean moment of inertia as defined by Jeffreys).

(2) The model should be physically plausible. All hypotheses should be explicitly stated, and reasons given for their adoption, as far as possible.
(3) The model should be simple: in general an attempt should be made to keep the number of parameters needed to describe the model as low as possible consistent with (2).
(4) The model may be spherically symmetrical. Authors of a set of unsymmetrical models (e.g. oceanic, continental, shield, ...) are urged to present also their own version of an averaged spherically symmetric Earth.
(5) The model should be shown to be reasonably consistent with data on eigenfrequencies, dispersion curves and travel-times.
(6) If c_P, c_S and ρ of the model are specified at discrete levels, specific methods of interpolation should be indicated so that values may be unambiguously constructed for all points.'

Subsequently, at the Durham Assembly of the International Association of Seismology and Physics of the Earth's Interior in 1977, it was decided – scientific progress having outrun the Committee –

that the reference model should include also the specification of quality factors $Q_P(r)$ and $Q_S(r)$. By that date it had been discovered that correct allowance for the effects of attenuation on dispersion (see § 10.3) removed most of the discrepancy between travel-times, dispersion curves and eigenfrequencies.

The construction of a reference model was then entrusted to Professor Don L. Anderson and Professor Adam M. Dziewonski, working in cooperation. They reported back to the Committee in 1979 and were requested to publish their final proposal during 1980. Thus the earlier exploration of travel-time studies and the recent investigations of free oscillations, which have formed the subject to this book, move towards an agreed Earth-model. Since direct observations extend downwards only a few kilometres into the Earth, this marks a splendid triumph of the scientific method.

APPENDIX

A.1 Some relations for spherical polar coordinates

Relations between the rectangular Cartesian coordinates (x, y, z) and the spherical polar coordinates (r, ϑ, φ) are (Fig. 2.3)

$$x = r \sin \vartheta \cos \varphi, \quad y = r \sin \vartheta \sin \varphi, \quad z = r \cos \vartheta. \qquad \text{(A.1.1)}$$

If ψ is a scalar function of r, ϑ and φ, and a vector $\mathbf{A}$ has components (R, Θ, Φ) which are scalar functions of r, ϑ and φ, we have (Morse & Feshbach, 1953, Chapter 1, tables)

$$\operatorname{grad} \psi = \left(\partial_r, \frac{1}{r}\partial_\vartheta, \frac{1}{r \sin \vartheta}\partial_\varphi \right)\psi, \qquad \text{(A.1.2)}$$

$$\operatorname{div} \mathbf{A} = \frac{1}{r^2}\partial_r(r^2 R) + \frac{1}{r \sin \vartheta}\partial_\vartheta(\Theta \sin \vartheta) + \frac{1}{r \sin \vartheta}\partial_\varphi \Phi, \qquad \text{(A.1.3)}$$

$$\nabla^2 \psi = \partial_r^2 \psi + \frac{2}{r}\partial_r \psi + \frac{1}{r^2}\left(\partial_\vartheta^2 \psi + \cot \vartheta \, \partial_\vartheta \psi + \frac{1}{\sin^2 \vartheta}\partial_\varphi^2 \psi \right), \qquad \text{(A.1.4)}$$

$$\operatorname{curl} \mathbf{A} = \left[\frac{1}{r \sin \vartheta}\{\partial_\vartheta(\Phi \sin \vartheta) - \partial_\varphi \Theta\}, \frac{1}{r}\left\{ \frac{1}{\sin \vartheta}\partial_\varphi R - \partial_r(r\Phi) \right\}, \frac{1}{r}\{\partial_r(r\Theta) - \partial_\vartheta R\} \right]. \qquad \text{(A.1.5)}$$

A.2 Derivation of equations (2.3.5) and (5.1.1)

We construct expressions for strain components and equations of motion in spherical polar coordinates (r, ϑ, φ) from those in rectangular coordinates.

The relations between the two sets of coordinates are

$$x = r \sin \vartheta \cos \varphi, \quad y = r \sin \vartheta \sin \varphi, \quad z = r \cos \vartheta.$$

The metric tensors g_{ij}, g^{ij} and Christoffel symbols Γ^i_{jk} for spherical polar coordinates are

$$\left.\begin{array}{lll} g_{11} = 1, & g_{22} = r^2, & g_{33} = r^2 \sin^2 \vartheta, \\ g^{11} = 1, & g^{22} = r^{-2}, & g^{33} = (r \sin \vartheta)^{-2}, \end{array}\right\} \tag{A.2.1}$$

if $i \neq j$, $g_{ij} = 0$ and $g^{ij} = 0$ (by orthogonality). Hence

$$\left.\begin{array}{ll} \Gamma^1_{22} = -r, & \Gamma^1_{33} = -r \sin^2 \vartheta, \\ \Gamma^2_{12} = \Gamma^2_{21} = r^{-1}, & \Gamma^2_{33} = -\sin \vartheta \cos \vartheta, \\ \Gamma^3_{13} = \Gamma^3_{31} = r^{-1}, & \Gamma^3_{23} = \Gamma^3_{32} = \cot \vartheta, \end{array}\right\} \tag{A.2.2}$$

all other Γ^i_{jk} being zero.

Since for orthogonal curvilinear coordinates it is not necessary to distinguish between covariant and contravariant tensors, we use lower indices. Differentiation of a first-order tensor follows the rule (Tyldesley, 1975, § 5.6)

$$T_i|_j = \partial_j T_i - \Gamma^\sigma_{ij} T_\sigma . \tag{A.2.3}$$

When $x_1 = r, x_2 = \vartheta$ and $x_3 = \varphi$ we obtain strain components derived from the tensor displacement T_i:

$$\left.\begin{array}{l} \varepsilon_{rr} = T_1|_1 = \partial_r T_r, \\ \varepsilon_{r\vartheta} = \tfrac{1}{2}(T_1|_2 + T_2|_1) = \tfrac{1}{2}(\partial_\vartheta T_r + \partial_r T_\vartheta) - T_\vartheta/r, \\ \varepsilon_{r\varphi} = \tfrac{1}{2}(T_1|_3 + T_3|_1) = \tfrac{1}{2}(\partial_r T_\varphi + \partial_\varphi T_r) - T_\varphi/r, \\ \vdots \end{array}\right\} \tag{A.2.4}$$

Denoting the *physical components* of the tensors by u_i and e_{ij}, we have

$$u_i = (g^{ii})^{1/2} T_i, \qquad e_{ij} = (g^{ii} g^{jj})^{1/2} \varepsilon_{ij}. \tag{A.2.5}$$

Putting (A.2.5) into (A.2.4) we obtain (2.3.5).

Differentiation of a second-order tensor follows the rule

$$T_{ij}|_\gamma = \partial_\gamma T_{ij} + \Gamma^i_{\sigma\gamma} T_{\sigma j} + \Gamma^j_{\sigma\gamma} T_{i\sigma} . \tag{A.2.6}$$

Hence we have

$$\left.\begin{array}{l} T_{11}|_1 = \partial_r T_{11}, \\ T_{12}|_2 = \partial_\vartheta T_{12} - r T_{22} + (1/r) T_{11}, \\ T_{13}|_3 = \partial_\varphi T_{13} - r \sin^2 \vartheta T_{33} + (1/r) T_{11} + \cot \vartheta T_{12}, \\ \vdots \end{array}\right\} \tag{A.2.7}$$

Denoting the physical stress components of T_{ij} by τ_{ij},

$$\tau_{ij} = (g_{ii} g_{jj})^{1/2} T_{ij}. \tag{A.2.8}$$

Hence, using (A.2.1), we have

$$\left.\begin{aligned} \tau_{rr} &= T_{11}, & \tau_{\vartheta\vartheta} &= r^2 T_{22}, & \tau_{\varphi\varphi} &= r^2 \sin^2 \vartheta T_{33}, \\ \tau_{r\vartheta} &= r T_{12}, & \tau_{r\varphi} &= r \sin \vartheta T_{13}, & \tau_{\vartheta\varphi} &= r^2 \sin \vartheta T_{23}. \end{aligned}\right\} \tag{A.2.9}$$

Using (A.2.7), we obtain from the first equation of (2.2.1)

$$\begin{aligned} \partial_r T_{11} + \partial_\vartheta T_{12} + \partial_\varphi T_{13} + (2/r) T_{11} - r T_{22} - r \sin^2 \vartheta T_{33} \\ + \cot \vartheta T_{12} + F_r = \rho \partial_t^2 u. \end{aligned}$$

Inserting (A.2.8) into this equation we get the first equation of (5.1.1) as

$$\begin{aligned} &\partial_r \tau_{rr} + \frac{1}{r} \partial_\vartheta \tau_{r\vartheta} + \frac{1}{r \sin \vartheta} \partial_\varphi \tau_{r\varphi} \\ &\qquad + \frac{1}{r}(2\tau_{rr} - \tau_{\vartheta\vartheta} - \tau_{\varphi\varphi} + \cot \vartheta \tau_{r\vartheta}) + F_r = \rho \partial_t^2 u. \end{aligned}$$

Similarly we get the second and third equations of (5.1.1) from corresponding equations of (2.2.1).

A.3 Continuity of gravitational potential and its normal gradient at a surface of discontinuity of density

Let the interface S between regions 1 and 2 (Fig. A.1) be smooth, and let density change discontinuously across S, but not within either region. Consider a region on S small enough to be considered as a plane circular disc $\mathscr{D}$ of area A. Imagine a penny-shaped cylinder $\mathscr{C}$ with plane faces parallel to $\mathscr{D}$ and generators normal to $\mathscr{D}$. We take this cylinder so small that to a sufficient approximation we can consider the part of it in each region to be of uniform density. Let the contributions to gravitational potential ϕ and force $\mathbf{F}$ at a point on the axis of $\mathscr{D}$ within $\mathscr{C}$ be ϕ_0, $\mathbf{F}_0$ from the material outside $\mathscr{C}$, and ϕ_1, $\mathbf{F}_1$ from material inside $\mathscr{C}$.

If we construct ϕ_0 and $\mathbf{F}_0$ as integrals we can easily prove that ϕ_0 and $\mathbf{F}_0$ are continuous across $\mathscr{D}$.

Next suppose that the radius of each face of the penny-shaped cylinder is a and the thickness h, and let h be small compared with a.

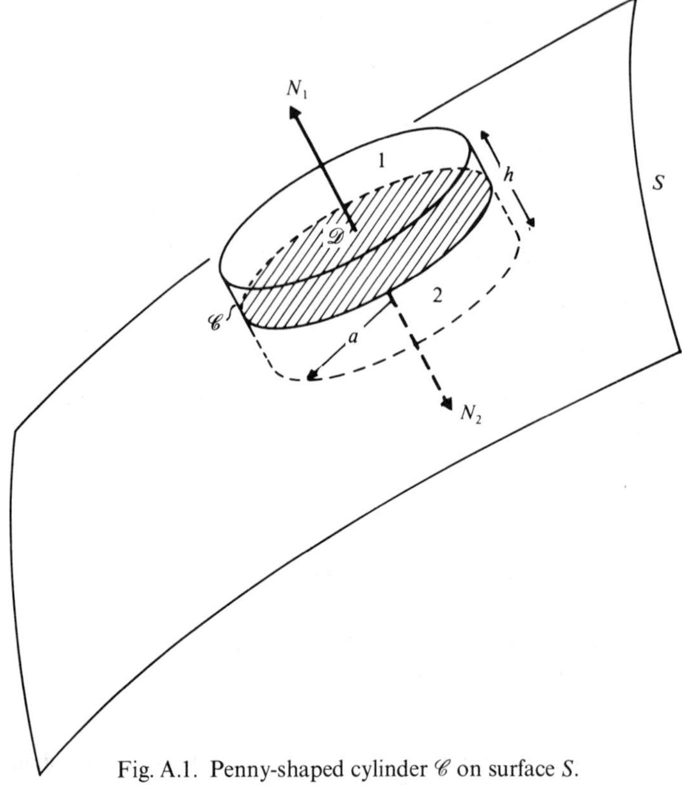

Fig. A.1. Penny-shaped cylinder $\mathscr{C}$ on surface S.

We now apply Gauss' formula. Let $\mathscr{F}_1$ be the flux of gravitational force outward across the face of $\mathscr{C}$ in region 1, and $\mathscr{F}_2$ that in region 2. Let $\mathscr{F}_3$ be the flux of gravitational force outwards across the curved boundary of $\mathscr{C}$. Then

$$\mathscr{F}_1 + \mathscr{F}_2 + \mathscr{F}_3 = -4\pi Gm, \tag{A.3.1}$$

where m is the mass of material within $\mathscr{C}$. $\mathscr{F}_3$ and m depend on h and can be made small compared with $\mathscr{F}_1$ and $\mathscr{F}_2$ by taking h/a small. Thus

$$\mathscr{F}_1 + \mathscr{F}_2 = O(Ah/a). \tag{A.3.2}$$

But by the choice of small enough a we can ensure that

$$\mathscr{F}_1 = N_1 A + \text{higher order terms},$$
$$\mathscr{F}_2 = N_2 A + \text{higher order terms},$$

where N_1 is the normal component of force (outwards) at the centre of the face of $\mathscr{C}$ in region 1, and N_2 similarly in region 2. Hence

$$(N_1 + N_2)A = O(Ah/a), \tag{A.3.3}$$

and $N_2 \to -N_1$ as $h \to 0$. Moreover, by symmetry, $\mathbf{F}_1$ has no component perpendicular to the axis of $\mathscr{C}$, to our order of approximation. Hence $\mathbf{F}_1$ is continuous across $\mathscr{D}$. We conclude that there is no discontinuity in $\mathbf{F}$ in crossing S, provided that our assumptions as to the nature of S and the density distributions are true.

We now turn to the potential. If we calculate the potential due to a uniform cylinder at a point on its axis, we find that there is no discontinuity at the end face. Thus neither part of the cylinder contributes a discontinuity to ϕ_1, and so, by consideration of the limit of $\phi = \phi_0 + \phi_1$ as a and h tend to zero, the potential is continuous across the interface.

A.4 Construction of solution (5.2.11)

Differentiating the product of the second equation of (5.2.4) and r, and subtracting the first equation, we obtain

$$(\mu/\rho_0)\mathscr{L}^2 H + \omega^2 H = \Gamma\Delta, \tag{A.4.1}$$

where

$$H = \partial_r V_l + (V_l - U_l)/r, \tag{A.4.2}$$

and $\mathscr{L}^2$ is the operator defined by

$$\mathscr{L}^2 \equiv \frac{\mathrm{d}^2}{\mathrm{d}r^2} + \frac{2}{r}\frac{\mathrm{d}}{\mathrm{d}r} - \frac{l(l+1)}{r^2}. \tag{A.4.3}$$

If we differentiate the first equation of (5.2.4) and add multiples of the first and second equations so as to give $\omega^2\Delta$, and use the third equation of (5.2.4), we obtain

$$\{(\lambda + 2\mu)/\rho_0\}\mathscr{L}^2\Delta + (\omega^2 + 4\Gamma)\Delta = l(l+1)\Gamma H. \tag{A.4.4}$$

From (A.4.1) and (A.4.4), we get

$$\{\omega^2 + (\mu/\rho_0)\mathscr{L}^2\}\{(\omega^2 + 4\Gamma)\Delta + (1/\rho_0)(\lambda + 2\mu)\mathscr{L}^2\Delta\} - l(l+1)\Gamma^2\Delta = 0. \tag{A.4.5}$$

H also satisfies this fourth-order equation. Using the definitions in (5.2.12), we write for (A.4.5)

$$(\mathscr{L}^2 + p^2)(\mathscr{L}^2 + q^2)\Delta = 0, \tag{A.4.6}$$

with solutions

$$\Delta = Z_l(p\zeta), \qquad Z_l(q\zeta). \tag{A.4.7}$$

Substituting for Δ in (A.4.1) and the third equation of (5.2.4), we find

$$H = \frac{\Gamma\beta^2 a^2}{\beta^2 a^2\omega^2 - p^2} Z_l(p\zeta), \qquad \frac{\Gamma\beta^2 a^2}{\beta^2 a^2\omega^2 - q^2} Z_l(q\zeta), \tag{A.4.8}$$

$$Y = -\frac{3a^2\Gamma}{p^2} Z_l(p\zeta), \qquad -\frac{3a^2\Gamma}{q^2} Z_l(q\zeta). \tag{A.4.9}$$

Since Δ, H and Y are now known, we can obtain U and V by solving (A.4.2) and (5.2.5) as a pair of simultaneous equations. The terms with constant multipliers T, T' in (5.2.11) are obtained as complementary solutions by solving the last equation of (5.2.4) with $\Delta = 0$. Factors $\Gamma l - \omega^2$ and $-\Gamma(l+1) - \omega^2$ in $Y_l(\zeta)$ in (5.2.11) are chosen so that U, V and Y satisfy (5.2.4).

When Δ is given by (A.4.7), H is obtained from either (A.4.1) or (A.4.4). Since the results must be the same, we get

$$\frac{\omega^2 + 4\Gamma - q^2/(\alpha^2 a^2)}{l(l+1)\Gamma} = \frac{\Gamma}{\omega^2 - q^2/(\beta^2 a^2)}, \tag{A.4.10}$$

which is equivalent to one of the equations (5.2.12).

A.5 Derivation of (8.5.3)

McNabb *et al.* have shown (1976) that the asymptotic form of the frequency equation for a Sturm–Liouville system with discontinuous coefficients is obtained as the condition that the following set of homogeneous linear equations is consistent:

$$\alpha_1 = 0, \tag{A.5.1}$$

$$\left.\begin{aligned} &\alpha_j \sin\chi_j + \beta_j \cos\chi_j - \gamma_j \beta_{j+1} = 0 && \text{(A.5.2)} \\ &\alpha_j \cos\chi_j - \beta_j \sin\chi_j - (1/\gamma_j)\alpha_{j+1} = 0 && \text{(A.5.3)} \end{aligned}\right\} \; j = 1, 2, \ldots, m-1,$$

$$\alpha_m \cos\chi_m - \beta_m \sin\chi_m = 0, \tag{A.5.4}$$

where in our problem χ_j is defined as in §8.3, and

$$\gamma_j^2 = (\beta_j \rho_j)/(\beta_{j+1}\rho_{j+1}). \tag{A.5.5}$$

As in (8.5.2), we have

$$R_j = (1-\gamma_j^2)/(1+\gamma_j^2), \tag{A.5.6}$$

so that, if we neglect terms of order R_j^2,

$$\gamma_j \approx 1 - R_j, \qquad (1/\gamma_j) \approx 1 + R_j. \tag{A.5.7}$$

Inserting these into (A.5.2) and (A.5.3) and inverting the equations we get, to first order in R_j,

$$\begin{pmatrix}\alpha_j\\ \beta_j\end{pmatrix} = (M_j + R_j N_j)\begin{pmatrix}\alpha_{j+1}\\ \beta_{j+1}\end{pmatrix}, \tag{A.5.8}$$

where

$$M_j \equiv \begin{pmatrix}\cos\chi_j & \sin\chi_j\\ -\sin\chi_j & \cos\chi_j\end{pmatrix}, \qquad N_j \equiv \begin{pmatrix}\cos\chi_j & -\sin\chi_j\\ -\sin\chi_j & -\cos\chi_j\end{pmatrix}. \tag{A.5.9}$$

From (A.5.8) we have approximately, neglecting $O(R^2)$,

$$\begin{aligned}\begin{pmatrix}\alpha_1\\ \beta_1\end{pmatrix} &\approx (M_1 + R_1 N_1)(M_2 + R_2 N_2)\ldots(M_{m-1} + R_{m-1}N_{m-1})\begin{pmatrix}\alpha_m\\ \beta_m\end{pmatrix}\\ &\approx (M_1 M_1 \ldots M_{m-1} + R_1 N_1 M_2 \ldots M_{m-1}\\ &\quad + M_1 R_2 N_2 M_3 \ldots M_{m-1} + \ldots + M_1 M_2 \ldots R_{m-1} N_{m-1})\begin{pmatrix}\alpha_m\\ \beta_m\end{pmatrix}.\end{aligned} \tag{A.5.10}$$

If for M_j we write $M(\chi_j)$ and for N_j write $N(\chi_j)$, then we easily prove by matrix multiplication that

$$M_1 M_2 = M(\chi_1 + \chi_2), \tag{A.5.11}$$

$$M_1 N_2 = N(\chi_1 + \chi_2), \tag{A.5.12}$$

$$N_1 M_2 = N(\chi_1 - \chi_2). \tag{A.5.13}$$

Thus

$$M_1 M_2 \ldots M_{m-1} = M(\chi_1 + \chi_2 + \ldots + \chi_{m-1}), \tag{A.5.14}$$

$$M_1 N_2 \ldots M_{m-1} = N(\chi_1 + \chi_2 - \chi_3 \cdots - \chi_{m-1}), \tag{A.5.15}$$

$$M_1 M_2 \ldots N_{m-1} = N(\chi_1 + \chi_2 + \ldots + \chi_{m-1}). \tag{A.5.16}$$

Rewriting (A.5.10) as

$$\begin{pmatrix}\alpha_1\\ \beta_1\end{pmatrix}=\begin{pmatrix}P^{11} & P^{12}\\ P^{21} & P^{22}\end{pmatrix}\begin{pmatrix}\alpha_m\\ \beta_m\end{pmatrix}, \tag{A.5.17}$$

we get, from $\alpha_1=0$,

$$P^{11}\alpha_m+P^{12}\beta_m=0, \tag{A.5.18}$$

and from (A.5.4)

$$P^{11}\sin\chi_m+P^{12}\cos\chi_m=0. \tag{A.5.19}$$

But

$$P^{11}=\left\{M\left(\sum_1^{m-1}\chi_j\right)\right\}^{11} + \left\{\sum_{j=1}^{m-1}R_jN(\chi_1+\ldots+\chi_j-\chi_{j+1}-\cdots-\chi_{m-1})\right\}^{11}, \tag{A.5.20}$$

and similarly for P^{12}. Hence from (A.5.19)

$$\sin\chi_m\left\{\cos\phi+\sum_{j=1}^{m-1}R_j\cos\psi_j\right\} + \cos\chi_m\left\{\sin\phi+\sum_{j=1}^{m-1}R_j(-\sin\psi_j)\right\}=0, \tag{A.5.21}$$

where

$$\phi=\chi_1+\chi_2+\ldots+\chi_{m-1}, \tag{A.5.22}$$

$$\psi_j=\chi_1+\chi_2+\ldots+\chi_j-\chi_{j+1}-\cdots-\chi_{m-1}, \tag{A.5.23}$$

that is,

$$\sin(\chi_1+\chi_2+\ldots+\chi_m)-\sum_{j=1}^{m-1}R_j\sin(\psi_j-\chi_m)=0. \tag{A.5.24}$$

If we neglect R_j, our zero-order approximation is

$$\chi_1+\chi_2+\ldots+\chi_m=n\pi, \tag{A.5.25}$$

where n is any integer. Let the first-order approximation be

$$\chi_1+\chi_2+\ldots+\chi_m=n\pi+\delta, \tag{A.5.26}$$

so that

$$\delta=\omega\gamma-n\pi \quad \text{and } S=\delta/\pi. \tag{A.5.27}$$

Substituting (A.5.26) into (A.5.24) we get

$$(-1)^n \sin\delta = \sum_{j=1}^{m-1} R_j \sin\left(n\pi + \delta - 2\sum_{p=j+1}^{m} \chi_p\right). \qquad \text{(A.5.28)}$$

If we regard δ as small and neglect it in the expressions on the right-hand side, we obtain

$$S \approx -\frac{1}{\pi}\sum_{j=1}^{m-1} R_j \sin\left(2\sum_{p=j+1}^{m} \chi_p\right). \qquad \text{(A.5.29)}$$

Now writing $\chi_p = \omega\chi'_p$ and $\omega = n\pi/\gamma$ in the small terms, we get

$$S \approx -\frac{1}{\pi}\sum_{j=1}^{m-1} R_j \sin\left(\frac{2n\pi}{\gamma}\sum_{p=j+1}^{m} \chi'_p\right). \qquad \text{(A.5.30)}$$

A.6 Derivation of relations (9.1.5)

Taking the divergence and curl of **F**, we have

$$\left.\begin{aligned} \operatorname{div}\mathbf{F} &= \nabla^2\Psi, \\ \operatorname{curl}\mathbf{F} &= \operatorname{curl}\operatorname{curl}\mathbf{A} = -\nabla^2\mathbf{A}. \end{aligned}\right\} \qquad \text{(A.6.1)}$$

It is known that the solution of Poisson's equation

$$\nabla^2\chi = -q, \qquad \text{(A.6.2)}$$

is given by

$$\chi = \frac{1}{4\pi}\int \frac{q(x', y', z')}{R}\,\mathrm{d}V', \qquad \text{(A.6.3)}$$

where

$$R = \{(x - x')^2 + (y - y')^2 + (z - z')^2\}^{1/2}.$$

By consideration of the components in turn, we see that the solution of the vector Poisson's equation

$$\nabla^2\mathbf{X} = -\mathbf{q} \qquad \text{(A.6.4)}$$

is

$$\mathbf{X} = \frac{1}{4\pi}\int \frac{\mathbf{q}(x', y', z')}{R}\,\mathrm{d}V'. \qquad \text{(A.6.5)}$$

Applying this to (A.6.1) we find

$$\Psi = -\frac{1}{4\pi}\int \frac{1}{R}\operatorname{div}'\mathbf{F}\,\mathrm{d}V',$$

$$\mathbf{A} = \frac{1}{4\pi}\int \frac{1}{R}\operatorname{curl}'\mathbf{F}\,\mathrm{d}V'. \tag{A.6.6}$$

Denoting the outward normal to the bounding surface Σ' of a volume V' by $\hat{\mathbf{n}}$, we use Gauss's theorem

$$\int_{V'} \operatorname{div}\mathbf{X}\,\mathrm{d}V' = \int_{\Sigma'} \mathbf{X}\cdot\hat{\mathbf{n}}\,\mathrm{d}\Sigma', \tag{A.6.7}$$

and Stokes' theorem

$$\int_{V'} \operatorname{curl}\mathbf{X}\,\mathrm{d}V' = -\int_{\Sigma'} \mathbf{X}\wedge\hat{\mathbf{n}}\,\mathrm{d}\Sigma', \tag{A.6.8}$$

to obtain

$$\int_{V'} \operatorname{div}(\mathbf{F}/R)\,\mathrm{d}V' = \int_{V'} \{(\operatorname{div}\mathbf{F})/R + \mathbf{F}\cdot\operatorname{grad}(1/R)\}\,\mathrm{d}V'$$

$$= \int_{\Sigma'} \frac{1}{R}\mathbf{F}\cdot\hat{\mathbf{n}}\,\mathrm{d}\Sigma', \tag{A.6.9}$$

and

$$\int_{V'} \operatorname{curl}(\mathbf{F}/R)\,\mathrm{d}V' = \int_{V'} \{(\operatorname{curl}\mathbf{F})/R - \mathbf{F}\wedge\operatorname{grad}(1/R)\}\,\mathrm{d}V'$$

$$= -\int_{\Sigma'} \frac{1}{R}\mathbf{F}\wedge\hat{\mathbf{n}}\,\mathrm{d}\Sigma'. \tag{A.6.10}$$

We assume that the region in which **F** operates is bounded, and take Σ' so large that **F** is zero on Σ', and obtain (9.1.5).

REFERENCES

Alsop, L. E. (1963). Free vibrations of the Earth at very long period. *Bull. Seismol. Soc. Amer.* **53**, 483–502.

Alterman, Z. S., & Aboudi, D. (1969). Seismic pulse in a layered sphere: normal modes and surface waves. *J. Geophys. Res.* **74**, 2618–36.

Alterman, Z. S., Eyal, Y., & Merzer, A.M. (1974). On free oscillations of the Earth. *Geophys. Surveys* **1**, 409–28.

Alterman, Z., Jarosch, H., & Pekeris, C. L. (1959). Oscillations of the Earth. *Proc. Roy. Soc.* A **252**, 80–95.

Anderssen, R. S. (1977). The effect of discontinuities in density and shear velocity on the asymptotic overtone structure of torsional eigenfrequencies of the Earth. *Geophys. J. Roy. Astr. Soc.* **50**, 303–9.

Anderssen, R. S., & Cleary, J. R. (1974). Asymptotic structure in torsional free oscillations of the Earth–I. Overtone structure. *Geophys. J. Roy. Astr. Soc.* **39**, 241–68.

Anderssen, R. S., Cleary, J. R., & Dziewonski, A. M. (1975). Asymptotic structure of eigenfrequencies of spheroidal normal modes of the Earth. *Geophys. J. Roy. Astr. Soc.* **43**, 1001–6.

Backus, G. E., & Gilbert, F. (1961). The rotational splitting of the free oscillations of the Earth. *Proc. Nat. Acad. Sci.* **47**, 362–71.

Backus, G. E., & Gilbert, F. (1967). Numerical applications of a formalism for geophysical inverse problem. *Geophys. J. Roy. Astr. Soc.* **13**, 247–76.

Backus, G. E., & Gilbert, F. (1968). The resolving power of gross Earth data. *Geophys. J. Roy. Astr. Soc.* **16**, 169–205.

Backus, G. E., & Gilbert, J. F. (1970). Uniqueness in the inversion of inaccurate gross Earth data. *Phil. Trans. Roy. Soc.* A **266**, 123–92.

Backus, G. E., & Mulcahy, M. (1976). Moment tensors and other phenomenological descriptions of seismic sources–I. Continuous displacements. *Geophys. J. Roy. Astr. Soc.* **46**, 341–61.

Bamford, D., & Crampin, S. (1977). Seismic anisotropy – the state of the art. *Geophys. J. Roy. Astr. Soc.* **49**, 1–8.

Benioff, H., Gutenberg, B., & Richter, C. F. (1954). Progress report, Seismological Laboratory, California Institute of Technology, 1953. *Trans. Amer. Geophys. Union.* **3**, 979–87.

Benioff, H., Press, F., & Smith, S. (1961). Excitation of the free oscillations of the Earth by earthquakes. *J. Geophys. Res.* **66**, 605–19.

Ben–Menahem, A. (1964). Mode–ray duality. *Bull Seismol. Soc. Amer.* **54**, 1315–21.

Brekhovskikh, L. N. (1960). Waves in layered media. Academic Press, London.

Brodskii, M., & Levshin, A. (1979). An asymptotic approach to the inversion of free oscillation data. *Geophys. J. Roy. Astr. Soc.* **58**, 631–54.

Bromwich, T. J. I'A. (1898–99). On the influence of gravity on elastic waves, and in particular, on the vibration of an elastic globe. *Proc. London Math. Soc.* **30**, 98–120.

Brune, J. N. (1964). Travel times, body waves and normal modes of the Earth. *Bull. Seismol. Soc. Amer.* **54**, 2099–128.

Brune, J. N. (1966). P and S travel times and spheroidal normal modes of a homogeneous sphere. *J. Geophys. Res.* **71**, 2959–65.

Brune, J. N., Nafe, J., & Alsop, L. (1961). Polar phase shift of surface waves on a sphere. *Bull. Seismol. Soc. Amer.* **51**, 247–57.

Bryan, G. H. (1889). The waves on a rotating liquid spheroid. *Phil. Trans. Roy. Soc.* A **180**, 187–219.

Bullard, E. C. (1954). The interior of the Earth. In *The Earth as a Planet*, ed. G. K. Kuiper, pp. 57–137. University of Chicago Press.

Bullen, K. E. (1959). *An Introduction to the Theory of Seismology*, 2nd edn. Cambridge University Press.

Bullen, K. E. (1974). Introductory remarks on standard Earth model. *Phys. Earth. Planet. Int.* **9**, 1–3.

Bullen, K. E., & Haddon, R. A. W. (1967). Derivation of an Earth model from free oscillation data. *Proc. Nat. Acad. Sci.* **58**, 846–52.

Burridge, R. (1976). *Some Mathematical Topics in Seismology.* Courant Institute of Mathematical Sciences, Lecture Notes, New York University.

Burridge, R., & Knopoff, L. (1964). Body force equivalents for seismic dislocations. *Bull. Seismol. Soc. Amer.* **54**, 1901–14.

Chree, C. (1889). The equations of an isotropic elastic solid in polar and cylindrical co-ordinates, their solution and application. *Trans. Camb. Phil. Soc.* **14**, 250–369.

Cleary, J. R., Osborne, M. R., & Anderssen, R. S. (1974). Asymptotic structure in torsional free oscillation data for the Earth. *Nature* **250**, 400–1.

Crampin, S. (1977). A review of the effects of anisotropic layering on the propagation of seismic waves. *Geophys. J. Roy. Astr. Soc.* **49**, 9–27.

Dahlen, F. A. (1968). The normal modes of a rotating, elliptical Earth. *Geophys. J. Roy. Astr. Soc.* **16**, 329–67.

Dahlen, F. A. (1969). The normal modes of a rotating, elliptical Earth – II. Near-resonance multiplet coupling. *Geophys. J. Roy. Astr. Soc.* **18**, 397–436.

Dahlen, F. A. (1972). Elastic dislocation theory for a self-gravitating elastic configuration with an initial static stress field. *Geophys. J. Roy. Astr. Soc.* **28**, 357–83.

Dahlen, F. A. (1974). Inference of the lateral heterogeneity of the Earth from the eigenfrequency spectrum: a linear inverse problem. *Geophys. J. Roy. Astr. Soc.* **38**, 143–67.

Dahlen, F. A. (1978). Excitation of the normal modes of a rotating earth model by an earthquake fault. *Geophys. J. Roy. Astr. Soc.* **54**, 1–9.

Derr, J. S. (1969a). Free oscillation observations through 1968. *Bull. Seismol. Soc. Amer.* **59**, 2079–99.

Derr, J. S. (1969b). Internal structure of the Earth inferred from free oscillations. *J. Geophys. Res.* **74**, 5202–20.

Dewey, J., & Byerly, P. (1969). The early history of seismometry (to 1900). *Bull. Seismol. Soc. Amer.* **59**, 183–227.

Dziewonski, A. M., & Gilbert, F. (1972). Observations of normal modes from 84 recordings of the Alaskan earthquake of 1964 March 28. *Geophys. J. Roy. Astr. Soc.* **27**, 393–446.

Dziewonski, A. M., & Gilbert, F. (1973). Observations of normal modes from 84 recordings of the Alaskan earthquake of 1964 March 28 – II. Further remarks based on new spheroidal overtone data. *Geophys. J. Roy. Astr. Soc.* **35**, 401–37.

Dziewonski, A. M., Hales, A. L., & Lapwood, E. R. (1975). Parametrically simple Earth models consistent with geophysical data. *Phys. Earth Planet. Int.* **10**, 12–48.

Fix, G. (1967). Asymptotic eigenvalues of Sturm–Liouville systems. *J. Math. Anal. Appl.* **19**, 519–25.

Gilbert, F. (1971). Inverse problems for the Earth's normal modes. *In* Mathematical problems in the geophysical sciences, 2. *Lectures in Applied Mathematics* vol. 14, ed. W. H. Reid, pp. 107–27. *Amer. Math. Soc.*

Gilbert, F. (1975). Some asymptotic properties of normal modes of the Earth. *Geophys. J. Roy. Astr. Soc.* **43**, 1007–11.

Gilbert, F. (1976). The representation of seismic displacements in terms of travelling waves. *Geophys. J. Roy. Astr. Soc.* **44**, 275–80.

Gilbert, F., & Backus, G. E. (1965). The rotational splitting of the free oscillations of the Earth, 2. *Rev. Geophys.* **3**, 1–9.

Gilbert, F., & Buland, R. P. (1977). Dissipation of elastic energy (abstract only). *EOS, Trans. Amer. Geophys. Un.* **58**, 440.

Gilbert, F., & Dziewonski, A. M. (1975). An application of normal mode theory to the retrieval of structural parameters and source mechanisms from seismic spectra. *Phil. Trans. Roy. Soc.* A **278**, 187–269.

Gilbert, F., & MacDonald, G. J. F. (1960). Free oscillations of the Earth, I. Toroidal oscillations. *J. Geophys. Res.* **65**, 675–93.

Hall, G., & Watt, J. M. (eds.) (1976). Modern numerical methods for ordinary differential equations. Clarendon Press: Oxford.

Haskell, N. A. (1953). The dispersion of surface waves on multilayered solid media. *Bull. Seismol. Soc. Amer.* **43**, 17–34.

Hornbeck, R. W. (1975). Numerical methods. Quantum Publishers, Inc.: New York.

Jeans, J. H. (1923). The propagation of earthquake waves. *Proc. Roy. Soc.* A **102**, 554–74.

Jeffreys, H. (1967). Radius of the Earth's core. *Nature* **215**, 1365–6.

Jeffreys, H. (1976). *The Earth*, 6th edn. Cambridge University Press.

Jeffreys, H., & Jeffreys, B. S. (1956). *Methods of Mathematical Physics*, 3rd edn. Cambridge University Press.

Jobert, N. (1956). Evaluation de la période d'oscillation d'une sphère elastique hétérogène, par application du principe de Rayleigh. *Comptes Rendus* **243**, 1230–2.

Jobert, N. (1957a). Sur la période propre des oscillations spheroidales de la terre. *Comptes Rendus* **244**, 921–2.

Jobert, N. (1957b). Sur la période propre des oscillations spheroidales de la terre. *Comptes Rendus* **245**, 1941–3.

Jobert, N. (1959). Calcul de la dispersion des ondes de Love de grande période à la surface de la Terre. *Comptes Rendus* **249**, 1014–6.

Jordan, T. H., & Anderson, D. L. (1974). Earth structure from free oscillations and travel times. *Geophys. J. Roy. Astr. Soc.* **36**, 411–59.

Kanamori, H., & Anderson, D. L. (1977). Importance of physical dispersion in surface wave and free oscillation problems – Review. *Rev. Geophys. Space Phys.* **15**, 105–12.

Kelvin, Lord (see Thomson, W.)

Kennett, B., & Nolet, G. (1979). The influence of upper mantle discontinuities on the toroidal free oscillations of the Earth. *Geophys. J. Roy. Astr. Soc.* **56**, 283–308.

Kennett, B. L. N., & Woodhouse, J. H. (1978). On high-frequency spheroidal modes and the structure of the upper mantle. *Geophys. J. Roy. Astr. Soc.* **55**, 333–50.

Knott, C. G. (1908). *The Physics of Earthquake Phenomena*. Clarendon Press: Oxford.

Lamb, H. (1882). On the vibration of an elastic sphere. *Proc. London Math. Soc.* **13**, 189–212.

Lamb, H. (1932). *Hydrodynamics*, 6th edn. Cambridge University Press.

Landisman, M., Satô, Y., & Nafe, J. (1965). Free vibrations of the Earth and the properties of its deep interior regions. Part 1: Density. *Geophys. J. Roy. Astr. Soc.* **9**, 439–502.

Lapwood, E. R. (1975). The effect of discontinuities in density and rigidity on torsional eigenfrequencies of the Earth. *Geophys. J. Roy. Astr. Soc.* **40**, 453–64.

Lapwood, E. R., & Sato, R. (1977). The asymptotic distribution of torsional eigenfrequencies of a spherical shell III. *J. Phys. Earth* **25**, 361–76.

Lapwood, E. R., & Sato, R. (1978). The pattern of eigenfrequencies of overtones of torsional oscillations of a layered spherical shell. *J. Comput. Phys.* **29**, (Alterman Memorial Issue), 412–20.

Lapwood, E. R., & Sato, R. (1979). The pattern of eigenfrequencies of radial overtones which is predicted for a specified Earth-model. *Annali di Geofisica*, **30**, 459–69.

Larmor, J. (1897). On the theory of the magnetic influence on spectra; and on the radiation from moving ions. *Phil. Mag.* (Math. and phys. papers II, Cambridge University Press 140–9.)

Liu, H. P., Anderson, D. L., & Kanamori, H. (1976). Velocity dispersion due to anelasticity; implications for seismology and mantle composition. *Geophys. J. Roy. Astr. Soc.* **47**, 41–58.

Love, A. E. H. (1911). *Some Problems of Geodynamics.* Cambridge University Press.

Love, A. E. H. (1927). *A Treatise on the Mathematical Theory of Elasticity*, 4th edn. Cambridge University Press.

Luh, P. C. (1973). Free oscillations of the laterally inhomogeneous Earth: quasi-degenerate multiplet coupling. *Geophys. J. Roy. Astr. Soc.* **32**, 187–202.

Luh, P. C. (1974). Normal modes of a rotating, self-gravitating inhomogeneous Earth. *Geophys. J. Roy. Astr. Soc.* **38**, 187–224.

Madariaga, R. (1972). Toroidal free oscillations of the laterally heterogeneous Earth. *Geophys. J. Roy. Astr. Soc.* **27**, 81–100.

Malvern, L. E. (1969). *Introduction to the Mechanics of a Continuous Medium.* Prentice Hall: New Jersey.

Matumoto, T. (1953). Transmission and reflection of seismic waves through a multilayered elastic medium. *Bull. Earthq. Res. Inst.* **31**, 261–73.

Matumoto, T., & Satô, Y. (1954). On the vibration of an elastic globe with one layer. The vibration of the first class. *Bull. Earthq. Res. Inst.* **32**, 247–58.

McNabb, A., Anderssen. R. S., & Lapwood, E. R. (1976). Asymptotic behaviour of the eigenvalues of a Sturm–Liouville system with discontinuous coefficients. *J. Math. Anal. Appl.* **54**, 741–51.

Milne, W. E. (1949). *Numerical Calculus.* Princeton University Press.

Morse, P. M., & Feshbach, H. (1953). *Methods of Theoretical Physics*, 2 vols. McGraw-Hill: New York.

Ness, N. F., Harrison, C. J., & Slichter, L. B. (1961). Observations of the free oscillations of the Earth. *J. Geophys. Res.* **66**, 621–9.

Nolet, G., & Kennett, B. (1978). Normal-mode representations of multiple-ray reflections in a spherical Earth. *Geophys. J. Roy. Astr. Soc.* **53**, 219–26.

Nowroozi, A. A. (1965). Eigenvibrations of the Earth after the Alaskan earthquake. *J. Geophys. Res.* **70**, 5145–56.

Odaka, T. (1978). Derivation of asymptotic frequency equations in terms of ray and normal mode theory and some related problems – radial and spheroidal oscillations of an elastic sphere. *J. Phys. Earth* **26**, 105–21.

Odaka, T., & Usami, T. (1978). Some properties of spheroidal modes of a homogeneous elastic sphere with special reference to radial dependence of displacement. *J. Comput. Phys.* **29** (Alterman Memorial Issue), 431–45.

Officer, C. B. (1951). Normal mode propagation in three-layered liquid half-space by ray theory. *Geophysics* **16**, 207–12.

Onda, I., & Sato, R. (1969). Transformation of an elastic wave solution related to transformation of the origin of the polar coordinate system. *Bull. Earthq. Res. Inst.* **47**, 599–611.

Parker, R. L. (1977). Understanding inverse theory. *Ann. Rev. Earth Planet. Sci.* **5**, 35–64.

Paschwitz, R. (1893). On the observation of earthquakes at great distances from the origin, with special relation to the great earthquake of Kumamoto, July 28th, 1889. *Seismol. J. Japan* **18**, 111–4.

Pekeris, C. L., Alterman, Z., & Jarosch, H. (1961a). Rotational multiplets in the spectrum of the Earth. *Phys. Rev.* **122**, 1692–700.

Pekeris, C. L., Alterman, Z., & Jarosch, H. (1961b). Comparison of theoretical with observed values of the periods of free oscillation of the Earth. *Proc. Nat. Acad. Sci.*, **47**, 91–8.

Pekeris, C. L., & Jarosch, H. (1958). The free oscillation of the Earth. In *Contributions in Geophysics*, ed. H. Benioff. pp. 171–92. Pergamon Press: London.

Press, F. (1964). Long-period waves and free oscillations of the Earth. In *Research in Geophysics*, vol. 2, ed. H. Odishaw, pp. 1–26. M.I.T. Press.

Ramsey, A. S. (1961). *An Introduction to the Theory of Newtonian Attraction.* Cambridge University Press.

Randall, M. J. (1976). Attenuative dispersion and frequency shifts of the Earth's free oscillations, *Phys. Earth Planet. Int.*, **12**, P1–P4.

Rayleigh, Lord (1896). *The Theory of Sound*, 2nd edn. Macmillan: London.

Sabatier, P. C. (1977). On geophysical inverse problems and constraints. *J. Geophys.* **43**, 115–37.

Sailor, R. V., & Dziewonski, A. M. (1978). Measurements and interpretation of normal mode attenuation. *Geophys. J. Roy. Astr. Soc.* **53**, 559–81.

Saito, M. (1967). Excitation of free oscillations and surface waves by a point source in a vertically heterogeneous Earth. *J. Geophys. Res.* **72**, 3689–99.

Sato, R., & Lapwood, E. R. (1977a). The asymptotic distribution of torsional eigenfrequencies of a spherical shell, I. *J. Phys. Earth.* **25**, 257–82.

Sato, R., & Lapwood, E. R. (1977b). The asymptotic distribution of torsional eigenfrequencies of a spherical shell, II. *J. Phys. Earth* **25**, 345–60.

Satô, Y. (1954). Study on surface waves XII. Non-dispersive surface waves. *Bull. Earthq. Res. Inst.* **32**, 349–60.

Satô, Y. (1964). Soft core spectrum splitting of the torsional oscillation of an elastic sphere and related problems. *Bull. Earthq. Res. Inst.* **42**, 1–10.

Satô, Y., Landisman, M., & Ewing, M. (1960). Love waves in a heterogeneous, spherical Earth. *J. Geophys. Res.* **65**, 2399–404.

Satô, Y., & Matumoto, T. (1961). Vibration of an elastic globe with a homogeneous mantle over a homogeneous core. Vibration of the first class. *J. Phys. Earth* **9**, 1–16.

Satô, Y., & Usami, T. (1962a). Basic study on the oscillation of a homogeneous elastic sphere. I. Frequency of the free oscillations. *Geophys. Mag.* **31**, 15–24.

Satô, Y., & Usami, T. (1962b). Basic study on the oscillation of a homogeneous elastic sphere. II. Distribution of displacement. *Geophys. Mag.* **31**, 25–47.

Satô, Y., & Usami, T. (1964). Propagation of spheroidal disturbances on an elastic sphere with a homogeneous mantle and a core. *Bull. Earthq. Res. Inst.* **42**, 407–25.

Satô, Y., Usami, T., & Landisman, M. (1968). Theoretical seismograms of torsional disturbances excited at a focus within a heterogeneous spherical Earth – Case of a Gutenberg–Bullen A′ Earth-model. *Bull. Seismol. Soc. Amer.* **58**, 133–70.

Satô, Y., Usami, T., Landisman, M., & Ewing, M. (1963). Basic study on the oscillation of a sphere. Part V: Propagation of torsional disturbances on a radially heterogeneous sphere. Case of a homogeneous mantle with a liquid core. *Geophys. J. Roy. Astr. Soc.* **8**, 44–63.

Scarborough, J. B. (1962). *Numerical Mathematical Analysis*, 5th edn. Clarendon Press: Oxford.

Shida, T. (1925). On the possibility of observing the free vibrations of the Earth. *Nagaoka Anniversary Volume*, 109–20.

Singh, S. J., & Ben-Menahem, A. (1969a). Eigenvibrations of the Earth excited by finite dislocations – I Toroidal oscillations. *Geophys. J. Roy. Astr. Soc.* **17**, 151–77.

Singh, S. J., & Ben-Menahem, A. (1969b). Eigenvibrations of the Earth excited by finite dislocations – II Spheroidal oscillations. *Geophys. J. Roy. Astr. Soc.* **17**, 333–50.

Slichter, L. B. (1967). Spheroidal oscillations of the Earth. *Geophys. J. Roy. Astr. Soc.* **14**, 171–7.

Smith, S. W. (1966). Free oscillations excited by the Alaskan earthquake. *J. Geophys. Res.* **71**, 1183–93.

Sokolnikoff, I. S. (1956). *Mathematical Theory of Elasticity*, 2nd edn. McGraw-Hill: New York.

Stoneley, R. (1924). The elastic waves at the surface of separation of two solids. *Proc. Roy. Soc.* A **106**, 416–28.

Stoneley, R. (1926). The elastic yielding of the Earth. *M.N.R.A.S. Geophys. Suppl.* **1**, 356–9.

Stoneley, R. (1961). The oscillations of the Earth. *Physics and Chemistry of the Earth*, vol. 4, pp. 239–50. Pergamon Press: London.

Takeuchi, H. (1959). Torsional oscillations of the Earth and some related problems. *Geophys. J.* **2**, 89–100.

Takeuchi, H., & Saito, M. (1972). Seismic surface waves. In *Methods in Computational Physics*, vol. 11, ed. B. A. Bolt *et al.*, pp. 217–95. Academic Press: New York.

Thomson, W. (Lord Kelvin) (1863a). On the rigidity of the Earth. *Phil. Trans. Roy. Soc.* **153**, 573–82.

Thomson, W. (Lord Kelvin) (1863b). Dynamical problems regarding elastic spheroidal shells and spheroids of incompressible liquid. *Phil. Trans. Roy. Soc.* **153**, 583–616.

Thomson, W. T. (1950). Transmission of elastic waves through a stratified solid medium. *J. Appl. Phys.* **21**, 89–93.

Tyldesley, J. R. (1975). *An Introduction to Tensor Analysis for Engineers and Applied Scientists.* Longman: London.

Usami, T. (1962). Some remarks on the solutions of the equation of motion for a homogeneous and isotropic elastic medium. *Geophys. Mag.* **31**, 1–13.

Usami, T., Kano, K., & Satô, Y. (1962). Some remarks on the solutions of the equation of motion for a homogeneous and isotropic elastic medium. II. Relation between various solutions of rotational type. *Geophys. Mag.* **31**, 623–32.

Usami, T., Kotake, Y., & Satô, Y. (1967). Soft core spectrum splitting and related problems of the spheroidal oscillations of an elastic sphere with a homogeneous mantle and core. *Bull. Earthq. Res. Inst.* **45**, 945–62.

Usami, T., Odaka, T., & Satô, Y. (1970). Theoretical seismograms and earthquake mechanism. Part I. Basic principles. Part II. Effect of time function on surface waves. *Bull. Earthq. Res. Inst.* **48**, 533–79.

Usami, T., & Satô, Y. (1966). Theoretical seismograms of spheroidal type on the surface of a homogeneous gravitating spherical Earth. *Bull. Earthq. Res. Inst.* **44**, 779–91.

Usami, T., & Satô, Y. (1972). Theoretical seismograms – introductory explanation and basic concept. *Earth-Science Reviews* **8**, 291–301.

Usami, T., Satô, Y., & Landisman, M. (1966). Preliminary study of the propagation of spheroidal disturbances on the surface of a heterogeneous spherical Earth. *Geophys. J. Roy. Astr. Soc.* **11**, 243–51.

Usami, T., Satô, Y., Landisman, M., & Odaka, T. (1968). Theoretical seismograms of spheroidal type on the surface of a gravitating elastic sphere. III. Case of a homogeneous mantle with a liquid core. *Bull. Earthq. Res. Inst.* **46**, 791–819.

Woodhouse, J. H. (1977). Oscillations of the Earth. Lectures given at the International Centre for Theoretical Physics, Trieste (unpublished).

Woodhouse, J. H., & Dahlen, F. A. (1978). The effect of a general aspherical perturbation on the free oscillations of the Earth. *Geophys. J. Roy. Astr. Soc.* **53**, 335–54.

Zharkov, B. N., & Lyubimov, V. M. (1970a). Torsional oscillations of a spherically asymmetrical model of the Earth. *Bull. (Izv.) Akad. Sci. USSR, Phys. Solid Earth* **2**, 71–6.

Zharkov, B. N., & Lyubimov, V. M. (1970b). Theory of spheroidal vibrations for a spherically asymmetric model of the Earth, *Bull. (Izv.) Akad. Sci. USSR, Phys. Solid Earth* **10**, 613–8.

NOTATION

The following list contains only symbols used frequently and systematically throughout this book. Each symbol is followed by:

(a) the number of the section where it is introduced;
(b) the number of the equation where it is defined or first used; and
(c) a brief description.

Subscripts and superscripts are omitted where no ambiguity arises. Vectors and tensors are represented by bold face characters.

Symbol	*Section*	*Definition or first use*	*Description*
a	2.3	(2.3.2)	radius of sphere or of the Earth
a_i	6.1	(6.1.3)	parameter in variational method
$a_k(t)$	9.3.2	(9.3.12)	parameter in eigenfunction expansion
a_p	10.1.2	(10.1.2)	parameter in linear combination of data kernels
$\mathbf{A}$	9.1.1	(9.1.2)	vector potential for **F**
$\mathcal{A}, \mathcal{B}, \mathcal{C}$	6.1.3	(6.1.8,11)	matrices in Thomson–Haskell method
b	3.1	(3.1.1)	radius of Earth's core
$\mathbf{B}, \mathbf{C}$	2.2	(2.2.8, 18)	vector potentials for **u**
c	1.2	(2.5.7)	phase-velocity (general)
c_g	2.5	(2.5.9)	group-velocity
c_P, c_S	2.2	(2.2.14, 17)	P and S velocities

c_{Rayleigh}	2.5	(2.5.19)	velocity of Rayleigh wave on surface of half-space
C_{ij}	6.1	(6.1.15)	elements of matrix $\mathscr{C}$
c_{ijkl}	9.3.1	(9.3.3)	elastic constants
E_{pq}	10.1.3	(10.1.17)	estimated variance matrix of errors
E_l^t, E_l^S, E_l^T	9.1.4	(9.1.24)	surface stress components
e_{ij}	2.2	(2.2.5)	strain tensor
$\left.\begin{matrix} e_{rr}, e_{\vartheta\vartheta}, e_{\varphi\varphi} \\ e_{r\vartheta}, e_{\vartheta\varphi}, e_{\varphi r} \end{matrix}\right\}$	2.3	(2.3.5)	strain components
$F_i, F_r, F_\vartheta, F_\varphi$	2.2	(2.2.1), (5.1.1)	components of body force
$f(t), \bar{f}(\omega)$	9.1.5	(9.1.6,27)	time-variation of applied force and its Fourier transform
$\mathbf{F}$	5.1	(5.1.11)	body force
$\mathbf{F}_0$	9.1	(9.1.6)	point force
G	5.1	(5.1.3)	gravitational constant
g	5.1	(5.1.2)	gravitational acceleration
$G_p(r)$	10.1.1	(10.1.1)	data kernel
h	2.2	(2.2.13)	wave number for P wave
$h_m^{(2)}$	9.1.1	(9.1.13)	spherical Bessel function
j_l	2.2		spherical Bessel function of the first kind
k	(1) 2.2 (2) 1.2	(2.2.17)	wave number for S wave mode number
$(\mathscr{K}, \mathscr{L}, \mathscr{M})$	7.1	(7.1.6)	stress-gradients
L	2.5	(2.5.5)	wave-length
l	1.3		Legendre degree
$\mathbf{L}$	9.3.2	(9.3.9)	vector operator
m	1.3		longitudinal order number
M_{ij}	9.3.1	(9.3.7)	moment tensor
$\mathscr{M}_{ij}$	9.3.2		moment rate tensor
M_R, M_U	10.3	(10.3.2,6)	relaxed, unrelaxed modulus

N	(1) 6.1.3		number of shells
	(2) 9.3.3		number of observing stations
n	2.3	(2.3.15)	overtone number
$\hat{\mathbf{n}}$	10.2	(10.2.5)	unit normal
p	(1) 5.1	(5.1.2)	hydrostatic pressure
	(2) 8.3.2	(8.3.35)	ray parameter
$P_l^m(\cos\vartheta)$	1.3	(2.2.25)	Associated Legendre function
Q	1.9		quality factor
$q(s)$	8.1.2	(8.1.25)	coefficient in transformed Sturm–Liouville equation
Q_k	9.3.3		quality factor for vibration in kth mode
R	(1) 4.3		$c_{\mathrm{Si}}/c_{\mathrm{So}}$
	(2) 8.3	(8.3.20,25)	coefficient of reflection
	(3) 9.1.1	(9.1.4)	distance of field point from point of force
(r, ϑ, φ)	1.3		spherical polar coordinates
(R, Θ, Φ)	(1) 1.3		factors of eigenfunction
	(2) A.1		(r, ϑ, φ)-components of $\mathbf{A}$
$\left.\begin{matrix}\widehat{rr}\ \widehat{\vartheta\vartheta}\ \widehat{\varphi\varphi} \\ \widehat{r\vartheta}\ \widehat{\vartheta\varphi}\ \widehat{\varphi r}\end{matrix}\right\}$	2.3	(2.3.3)	stress components
$\mathbf{r}$	6.1.2	(6.1.2)	position vector of general point
S	(1) 5.1	(5.1.6)	surface bounding volume V
	(2) 8.3.1	(8.3.28)	measure of solotone effect
S, S_0	7.1, 9.1.1		reference frames
$S(r_0)$	10.1.2	(10.1.5)	spread about r_0
s	8.1.1	(8.1.12)	independent variable after Liouville transformation
${}_nS_l^m$	1.3		spheroidal mode of free oscillation
T	(1) 2.3	(2.3.12)	period of oscillation
	(2) 8.3.3	(8.3.40)	travel-time along a ray

	(3) 8.6.2	(8.6.7)	coefficient of transmission
t	1.2		time
$\mathbf{T}_0$	10.2	(10.2.2)	static stress tensor
$\tilde{\mathbf{T}}_0$	10.2	(10.2.9)	increment in P–K stress tensor
${}_nT_l^m$	1.3		toroidal mode of free oscillation
$(U(r), V, W)$	2.5	(2.5.1,2)	components of eigenfunction
(u, v, w)	2.2	(2.2.28–30)	displacement components, rectangular or polar
$u_i, \mathbf{u}$	2.2	(2.2.1,6)	displacement vector
$(\bar{u}, \bar{v}, \bar{w})$	9.1.4	(9.1.25)	displacement components in frequency domain
V	5.1	(5.1.6)	volume within surface S
V_s	8.3.2	(8.3.35)	speed of wave-front along surface
$x_i, (x, y, z)$	2.2		rectangular Cartesian coordinates
$Y(r)$	5.2	(5.2.3)	factor in disturbed gravitational potential
y_l	2.2		spherical Bessel function of the second kind
$y_1 \cdots y_6$	6.1	(6.1.21,26)	dependent variables for computations
$Z(s)$	8.1.2	(8.1.23)	dependent variable after Liouville transformation
z_l	2.2		general spherical Bessel function
$\hat{\mathbf{z}}$	7.1	(7.1.2)	unit vector in direction Oz
α	7.1	(7.1.9)	Ω/ω_N
Γ	5.2	(5.2.1)	the constant $\frac{4}{3}\pi G\rho_0$
γ	1.8	(3.4.1)	travel-time of body wave along radius

Γ_{ij}	9.3.1	(9.3.1)	stress glut
γ_p	10.1	(10.1.1)	gross Earth functional
${}_n\gamma_l$	8.2		estimate of γ
Δ	(1) 2.2	(2.2.4)	dilatation
	(2) 8.3.2	(8.3.35,40)	epicentral distance (angle)
ΔD	9.1.3	(9.1.21)	discontinuity at source surface
δ_{ij}	2.2	(2.2.2)	Kronecker delta
δ_i^{lm}	9.1.4	(9.1.25)	discontinuities at source surface
∂_n	5.1	(5.1.14)	derivative along normal
ε	(1) 7.2	(7.2.2)	ellipticity of the Earth
	(2) 10.3.1	(10.3.1)	strain
ε_{ij}	A.2	(A.2.4)	strain tensor
ξ	2.3	(2.3.14)	*ha*
η	2.3	(2.3.12)	*ka* (non-dimensional frequency)
ζ	3.1	(3.1.4)	*kb*
Λ	9.1.2	(9.1.19)	factor in Φ
$\boldsymbol{\Lambda}$	10.2		generalised elastic constant
λ, μ	1.3		Lamé's constants
ν	2.5	(2.5.12)	$l+\frac{1}{2}$
ρ	2.2	(2.2.1)	density
Σ	9.1.1	(9.1.8)	surface
σ	(1) 10.1.3		parameter in trade-off curve
	(2) 10.3.1	(10.3.1)	stress
σ_{ij}	9.3.1	(9.3.1)	physical stress tensor
$\tau(l)$	7.1	(7.1.18)	splitting coefficient
τ_{ij}	2.2	(2.2.1)	stress tensor
(θ, ϕ)	8.3.1	(8.3.24)	parameters in S(2)
Φ	9.1.1	(9.1.2)	scalar potential for $\mathbf{u}$
$\Phi(\vartheta, \varphi)$	9.2.1	(9.2.3,7)	angular dependence of external force

$\phi(t)$	10.3.1	(10.3.4)	creep function
(ϕ, χ, ψ)	2.2	(2.2.7,15,18)	Scalar potentials for **u**
χ	4.1	(4.1.6)	*hb*
χ_1, χ_2	8.3	(8.3.21)	phases in solotone formulae
$\mathbf{\Psi}$	9.1	(9.1.2)	scalar potential for **F**
ψ	5.1	(5.1.3)	gravitational potential
$\mathbf{\Omega}$	7.1	(7.1.1)	rotation vector of the Earth
ω	1.2		angular frequency
${}_n\omega_l^m$	1.3		angular frequency of an (l, m, n) mode

AUTHOR INDEX

SUBJECT INDEX